Michael Bauer

Raman spectroscopy of laser induced material alterations

Michael Bauer

Raman spectroscopy of laser induced material alterations

caused by the excitation laser (local heating) and surface damage by UV-lasers

Südwestdeutscher Verlag für Hochschulschriften

Imprint

Any brand names and product names mentioned in this book are subject to trademark, brand or patent protection and are trademarks or registered trademarks of their respective holders. The use of brand names, product names, common names, trade names, product descriptions etc. even without a particular marking in this work is in no way to be construed to mean that such names may be regarded as unrestricted in respect of trademark and brand protection legislation and could thus be used by anyone.

Publisher:
Südwestdeutscher Verlag für Hochschulschriften
is a trademark of
Dodo Books Indian Ocean Ltd., member of the OmniScriptum S.R.L Publishing group
str. A.Russo 15, of. 61, Chisinau-2068, Republic of Moldova Europe
Printed at: see last page
ISBN: 978-3-8381-1949-6

Zugl. / Approved by: München, Ludwig-Maximilians-Universität, Dissertation, 2010

Copyright © Michael Bauer
Copyright © 2010 Dodo Books Indian Ocean Ltd., member of the OmniScriptum S.R.L Publishing group

Contents

Table of Contents	ii
List of Figures	iii
List of Abbreviations	v
Kurzfassung	1
Abstract	2
1 Introduction	3
2 Raman spectroscopy	5
2.1 Interaction of light with matter	5
2.2 The Raman effect	6
2.3 Selection rules and Raman line determination	9
2.4 Influence of stress and temperature on Raman spectra	10
3 Methods	17
3.1 The confocal Raman microscope set up	17
3.1.1 Pinhole influence and light suppression	18
3.1.2 Gaussian laser focusing and beam parameters	19
3.2 Local heating due to the focused laser beam	21
3.3 Image generation	21
4 Spectroscopic characterisation	23
4.1 Silicon	23
4.1.1 Phonon dispersion curve and band structure	23
4.1.2 Raman scattering in silicon	25
4.1.3 Local heating	26
4.1.4 Experimental results	27
4.2 Silicon carbide	27

		4.2.1	Crystal structure and polytypes	28
		4.2.2	Phonon structure and mode folding	29
		4.2.3	Raman line shift	31
		4.2.4	Local heating	32
		4.2.5	Experimental results	33
	4.3	\multicolumn{2}{l}{Alteration of calcium fluoride caused by UV-light}	34
		4.3.1	Material properties of calcium fluoride	36
		4.3.2	Identification of kerogeneous carbon	37
		4.3.3	Experimental results	38
	4.4	\multicolumn{2}{l}{Titanomagnetites}	39

5 Summary 43

6 References 45

7 Publications 55

    7.1 Visualizing stress in silicon microcantilevers using scanning confocal Raman spectroscopy 55

    7.2 Temperature depending Raman line-shift of silicon carbide 61

    7.3 Nanoscale residual stress-field mapping around nanoindents in SiC by IR s-SNOM and confocal Raman microscopy 71

    7.4 Exterior surface damage of calcium fluoride out coupling mirrors for DUV lasers 81

    7.5 Onset of the optical damage in CaF_2 optics caused by deep-UV lasers 95

Appendix   I

Acknowledgements   XV

List of Figures

2.1	Typical configuration of IR and Raman spectroscopy	6
2.2	Jablonksi energy diagram of the scattering processes	7
3.1	Schematic of the confocal setup	17
3.2	Lateral resolution of a scattering point as a function of detector size	18
3.3	Illustrated Gaussian beam of a focused laser beam	20
3.4	Filter region for sum filtering and Lorentz fits	22
4.1	Electronic band structure and phonon dispersion curves of silicon	24
4.2	Refraction and extinction index of silicon with penetration depth and adsorption	24
4.3	Variation of peak centre and peak width due to local heating	26
4.4	Measurements of a silicon AFM cantilever	27
4.5	Side view ABC layer stacking and of silicon carbide	29
4.6	Phonon dispersion curve of silicon carbide	30
4.7	Mode folding of silicon carbide	30
4.8	Raman spectra of silicon carbide	31
4.9	Temperature-induced peak shift and increasing peak width of silicon and silicon carbide	33
4.10	Temperature-dependent shifting of peak centres and stress fields around a nanoindentation in SiC	33
4.11	Layer materials used for ultraviolet optical components	35
4.12	First- and second-order Raman spectra of kerogen carbon	37
4.13	Damage structure due to UV photon irradiation	39
4.14	Raman spectra of synthetic magnetite and titanomagnetite	40
4.15	Titanium concentration against laser power for beginning material alteration	41
4.16	Stripes of different titanomagnetites in a geologic sample	42

List of Abbreviations

AFM	atomic force microscope
ArF	argon fluoride
CaF_2	calcium fluoride
$CaCO_3$	calcium carbonate
CCD	charge coupled device
DUV	deep ultraviolet
FLA	folded longitudinal acoustic phonon
FLO	folded longitudinal optical phonon
FTA	folded transversal acoustic phonon
FTO	folded transversal optical phonon
FWHM	full width at half maximum
HR	high reflective (mirror)
IR	infrared
KrF	krypton fluoride
LA	longitudinal acoustic phonon
LO	longitudinal optical phonon
LOPC	LO-phonon plasmon coupled
MEMS	micro(electro) mechanical systems
MgF_2	magnesium fluoride
NA	numerical aperture
Nd:YAG	neodymium doped yttrium aluminium garnet
OC	out coupling (mirror)
PDP	phonon deformation potential
PSF	point spread function
Si	silicon
SiC	silicon carbide
SHG	second harmonic generation
s-SNOM	scattering scanning nearfield optical microscope
TA	transversal acoustic phonon
TEM	transverse electro magnetic (laser); or transmission electron microscope
TO	transversal optical phonon
UV	ultraviolet
VUV	vacuum ultraviolet

Kurzfassung

Die konfokale Raman-Spektroskopie ist eine leistungsfähige Methode zur Charakterisierung von Materialeigenschaften. Damit lassen sich mechanische Spannungen in homogenen Proben sowie die Verteilung von Probenbestandteilen optisch bestimmen. Zudem ist es mit der konfokalen Raman-Mikroskopie möglich Spannungsfelder in Silizium und Siliziumkarbid darzustellen, welche z.b. während mechanischer Belastung oder beim Herstellungsprozess entstehen. Diese Informationen sind für die Optimierung von Fertigungsprozessen und eine verbesserte Ausfallsicherheit von Mikrosystemen von Bedeutung. Für eine präzise Bestimmung von Spannungsfeldern ist eine genaue Untersuchung von thermischen Verschiebungen der Ramanlinien wichtig. Um die Auswirkung auf die Ramanspektren (thermische Linienverschiebung und -verbreiterung) zu ermitteln, wurden Silizium und unterschiedliche Siliziumkarbid Kristallmodifikationen kontrollierten Heizexperimenten unterzogen. Diese Kalibrierungen wurden benötigt, um thermische von mechanischen Effekten unterscheiden zu können, was an einem ausgelenkten Silizium-Cantilever gezeigt wurde. Zusätzlich zum äußeren Erwärmen kann es während der Messung zu einer lokalen Erhitzung der Proben durch den stark fokussierten Laserstrahl kommen. Für verlässliche Spannungsmessungen in mikrostrukturierten Silizium muss eine solche lokale Erwärmung berücksichtigt werden. Das Abbilden mittels Raman bietet hier die Möglichkeit Restspannungen sichtbar zu machen, die durch Eindrücke in die Oberfläche entstanden sind. Ein Vergleich zwischen Abbildungen eines Oberflächeneindruckes, welche mit einem Raman und einem optischen Streulicht-Nahfeldmikroskop (s-SNOM) aufgenommen wurden, deckte unterschiedliche Mechanismen bei der Bildentstehung auf. Während die Ramanstreuung Änderungen unterhalb der Oberfläche darstellen kann, ist die optische Nahfeldmikroskopie für oberflächennahe Verspannungen empfindlich.

Die Raman-Spektroskopie kann ebenfalls für die Bestimmung von Inhaltsstoffen in heterogenen Proben und die Charakterisierung von Materialveränderungen verwendet werden. Ein intensiver UV-Laser kann zum Beispiel eine Umwandlung von Kalziumfluorid eines Auskoppelspiegels zu Kalziumkarbonat (Kalzit) verursachen, was sich mit der konfokalen Raman Spektoskopie veranschaulichen ließ. Eine Veränderung des Probenmaterials kann aber auch durch den Anregungslaser selbst erfolgen. So traten oxidative Prozesse an Titanomagnetiten, eingelagert in geologische Proben auf, die durch den fokussierten Laser bedingte lokale Aufheizung hervorgerufen wurden. Diese laserinduzierte Veränderung wurde genauer an geologischen und synthetischen Titanomagnetiten untersucht.

Abstract

Confocal Raman spectroscopy is a powerful tool for material characterisation. It can be used for the optical characterisation of mechanical stress in homogeneous samples, for material identification, and for compositional mapping. In addition, confocal Raman microscopy is useful for the measurement of stress fields generated, for example, by mechanical loading or residual stress after fabrication in silicon and silicon carbide. This information is important for optimising manufacturing processes and increasing the reliability of micro devices. For precise stress measurements, the thermally induced shift of Raman peaks must be characterised in detail. For that purpose, silicon and different silicon carbide polytypes underwent controlled heating experiments to determine the effect of heating on Raman spectra (thermal peak shift and increasing line width). These calibrations are required to separate effects caused by temperature changes from those caused by mechanical stress. The separation of stress- and temperature-induced effects was demonstrated for the measurement of stress fields in a deflected silicon cantilever. In addition to external heating, the measurement itself may affect the sample temperature because the focused, intense light can locally heat the sample. Local heating must be taken into account to ensure reliable stress measurements in silicon micro-structures. Raman mapping may also be used to characterise residual stress fields caused by nanoindentation. A comparison between Raman mapping and scattering scanning near-field optical microscopy (s-SNOM) of a nanoindentation imprint in silicon carbide revealed subtle differences in the imaging mechanisms. Scattering-SNOM is sensitive to near surface stress, whereas Raman scattering can reveal sub-surface changes.

Raman spectroscopy can also be used for compositional mapping of heterogeneous specimens or for the characterisation of material alterations. For example, intense UV-laser light may induce the alteration from calcium fluorite to calcium carbonate (calcite) in outcoupling mirrors. These alterations can be visualised by confocal Raman microscopy. Material alterations can also be caused by the excitation laser itself. During the investigation of titanomagnetites that were embedded in geologic samples, oxidative processes occurred and they were driven by local heating due to the focused laser. The laser-induced alterations were investigated in detail on geologic and synthetic titanomagnetite samples.

1 Introduction

Compositional mapping or quantitative chemical analysis is important in micro- and nanostructured materials. In addition to other techniques, including atomic force microscopy (AFM) and scanning electron microscopy (SEM), Raman spectroscopy is becoming an increasingly common analysis method. It offers fast and contact-free measurements with easy sample preparation. Raman spectroscopy is a vibrational spectroscopy method that is based on the analysis of inelastically scattered light. The incident photons scatter off molecular vibrations (e.g., bending and rotational vibrations) or, in the case of crystals, off phonons (e.g., different modes of lattice vibration. As a result, specific spectral peak shifts occur, providing highly sensitive measurements of the material. The spectra consist of peaks that are unique in their position, shape and intensity. [1, 2] Usually, lasers are used as monochromatic excitation sources and the so-called Raman shift, i.e., the energy loss (or gain) of the scattered light, is given.

Raman spectroscopy can be used on a heterogeneous specimen to determine its composition. The spatial resolution is restricted only by the scattering volume. With a confocal set-up, the scattering volume can be controlled, allowing compositional mapping with sub-micrometer resolution. This sensitivity allows material composition questions to be answered with Raman spectroscopy. [1, 2]

In addition to sample composition, other parameters such as residual stress or temperature can be quantified with Raman spectroscopy. [3–5] Mechanical stress affects the phonon frequencies, which can be recognised as additional shifts of the Raman spectral peaks. With the continuing reduction of feature size in small devices such as micro-electro-mechanical systems (MEMS), stress fields in devices are becoming increasingly important. [3] One result of this continuing miniaturisation is the increase of mechanical stress in device structures. Different thermal expansion coefficients of different materials may induce mechanical stresses along structural edges and at corners during device operation. Monitoring these stresses is important to improve device reliability and to increase fabrication yield. In addition to mechanical stresses, temperature also affects a material's phonon scattering properties. The population of phonon modes increases with increasing temperature, and a variation in the population is expressed as a change in the peak width. Analysing the peak width allows one to measure the sample temperature optically.

Because confocal Raman spectroscopy uses a focused laser, the temperature in the scattering volume may increase significantly. Depending on the thermal conductivity and light reflectivity of the sample, it will be locally heated to some degree, which can affect the phonons directly. This local heating effect must be compensated for in stress measurements. The local laser heating may also induce material alterations or even induce the destruction of the sample. [6, 7]

In the thesis, the following issues will be discussed. The thesis starts in chapter 2 with the theoretical background of the Raman effect, explaining the selection rules for the scattering process. The influences of mechanical stress and varying temperature on Raman spectra are also briefly discussed. Chapter 3 describes the measurement system and explains the components and their influences on measurements. The spectroscopic characteristics of the materials are summarised in chapter 4. Important experimental results include the dependency of the spectra on mechanical stress and temperature in silicon and silicon carbide single crystals. Compositional mapping is shown for the failure analysis of calcium fluoride resonator mirrors impaired from high flux ultraviolet photons. Titanomagnetites particles, which are embedded in a deep sea basalt specimen, are a further example for chemical analysis. A concluding summary is given in chapter 5. The results that were published in scientific journals are provided in chapter 7.

2 Raman spectroscopy

Raman spectroscopy records the highly specific vibrational fingerprint of a material by analysing inelastically scattered light. Raman spectroscopy is based on the Raman effect, which was named after the Indian physicist Chandrasekhara Venkata Raman, who found in 1928 „*A new type of secondary radiation*". [8] Independently, the same observation was reported by Landside and Mandelbaum. [9] With the development of intense single mode lasers as powerful monochromatic light sources, the Raman effect became widely used in science and engineering. Confocal Raman spectroscopy has an intrinsic „nano-specificity"because it probes vibrations of chemical bonds and the propagation phonons in a small sample volume, typically less than 1 µm^3.

Vibrational stretching modes are highly characteristic for specific chemical bonds and often allow for compositional identification. In general, vibrational bending modes are more sensitive to the neighbouring entities and, hence, to the molecular short-range order. Librational and external modes, which depend on the structure, involve relative motions of entire units of molecules or atoms. Well-defined repetitive structures in crystals typically give rise to Raman spectra with sharp and well-resolved peaks. Spectra with broad bands indicate that different configurations or many electric or mass defects may exist in the crystal. Thus, if the proportion of atoms belonging to the (near-) surface region of the probed volume is significant, the disorder related to the various atomic arrangements contributes to the observed Raman features. [10]

2.1 Interaction of light with matter

Various processes can occur when light interacts with matter. In a case where the photon energy matches the difference between two real states of a molecule, the molecule can be excited into a higher electronic state. The excited molecule can relax to the ground state via a non-radiative process or via an additional luminous process (fluorescence). This process is used in absorption spectroscopy techniques such as infra-red (IR) spectroscopy. If the photon energy does not match the difference between the current state and another real state, it can be scattered inelastically (Brillouin and Raman scattering) or elastically (Rayleigh and Mie scattering). For an elastically scattered photon, the photon energy (wavelength) is the same before and after the scattering process. The process can be considered absorption of the photon (excitation of matter into a virtual intermediate state), which is followed by a prompt relaxation with re-emission of the photon. In contrast, inelastic Raman scattering is a nonlinear scattering process, involving quasiparticles such as phonons in a lattice or molecular vibrations (e.g., bending or rotation). [11–13] The amount of energy change (either loss or gain) by a photon is characteristic for the nature of vibration (phonon or chemical bond). Nevertheless, not all vibrations will be observable with Raman spectroscopy because Raman scattering depends on symmetry and polarisability. Figure 2.1 shows a schematic of typically used configurations for infrared and Raman spectroscopy. Whereas IR-spectroscopy analyses the absorbed light of a broadband infra-red light source, Raman analyses inelastically scattered laser light randomly emitted in each direction.

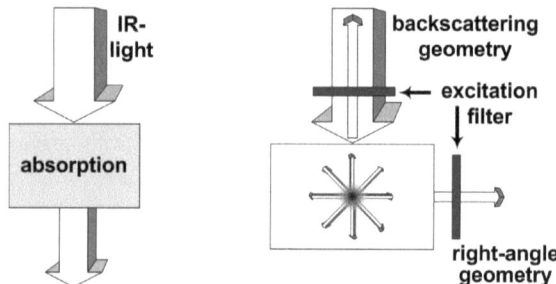

Figure 2.1: *Typical configuration of IR and Raman spectroscopy.*

2.2 The Raman effect

For a spectroscopic application of the Raman effect, the incident light has to be monochromatic. More than 99% of scattered light still has the same wavelength after scattering, and only a small amount of photons (10^{-6}%) are spectrally shifted with respect to the incident ones due to inelastic scattering. The Raman shift is given in (relative) wavenumbers with respect to the excitation. Despite blocking the unshifted photons, the spectra always show a peak representing the Rayleigh scattering together with the Raman shifted photons. These peaks are what give Raman spectra chemical and structural sensitivity. For Raman-scattered photons, the matter is excited into a virtual intermediate state and then relaxes to a vibrational state above the ground state. The photon transfers a certain amount of energy to

The Raman effect

the material, losing energy (Stokes shift). If the matter is already in a vibrationally excited state, inelastically scattered photons may also gain energy. The incident photon excites the material to a virtual intermediate state, which then can relax to its ground state. The scattered photon gains the energy difference, having a higher energy than before the scattering event (anti-Stokes shift). These three Raman scattering processes can be understood in a classical picture as a collision between the incident photon with matter. The Jablonski diagram in Fig. 2.2 shows the processes caused by incident photons. As in absorption spectroscopy (IR spectroscopy), the photon is absorbed and the system is excited to a higher state (Fig. 2.2(a)). If there is no state available that can be populated by the system, scattering occurs. Three important scattering processes include:

1. elastic scattering without energy transfer (Rayleigh scattering, 0.01%, Fig. 2.2(b)), the scattered photon has the same energy as before ($\hbar\omega_i = \hbar\omega_R$);

2. inelastic scattering with energy loss (Stokes shift, 10^{-6}%, Fig. 2.2(c)), the scattered photon loses energy ($\hbar\omega_i > \hbar\omega_s$); and

3. inelastic scattering with energy gain (anti-Stokes shift, 10^{-6}%, Fig. 2.2(d)), the scattered photon gains energy ($\hbar\omega_i < \hbar\omega_{as}$).

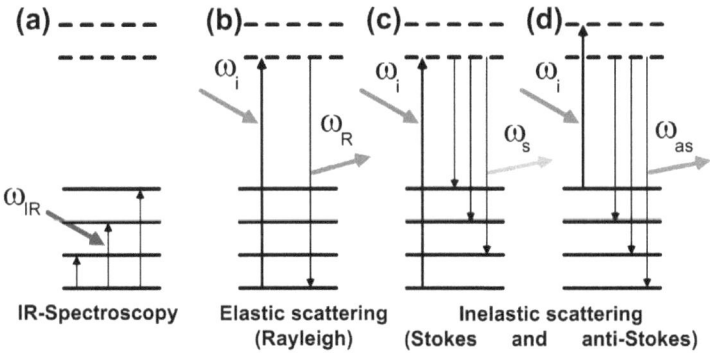

Figure 2.2: *Jablonski energy diagram of the scattering processes of light with Rayleigh scattering and Raman scattering (Stokes and anti-Stokes shifts). For comparison, an IR spectroscopy scattering scheme is shown on the left.*

Thus, Raman spectroscopy measures the energy that is transferred due to inelastic scattering. The energy transfer between both collision partners is equal to the difference between the two involved energy levels, $\hbar\omega_j$. With the energy $\hbar\omega_j$ transferred during the scattering process and the energy $\hbar\omega_i$ before scattering, the scattered photon will have the energy

$$\hbar\omega_{s/as} = \hbar\omega_i \mp \hbar\omega_j. \quad (2.1)$$

The plus and minus in equation 2.1 stand for the energy gain (anti-Stokes shift) or energy loss (Stokes) of the photon. The Stokes and anti-Stokes shifts are symmetric, determined by the difference of energy between the ground state and the vibrational state.

For scattering on molecules, the electric field of the photon affects the molecule's electrons, resulting in molecular distortion. The distorted molecule acquires a contribution to its dipole moment μ (even if initially non-polar):

$$\mu = \alpha E \quad (2.2)$$

where α is a proportional constant of molecule's polarisability. The polarisability corresponds to the ease with which the electron cloud around a molecule can be distorted. The induced dipole can emit or scatter light at the optical frequency of the incident light wave. Vibrational Raman scattering occurs because a molecular vibration can change the polarisability. The change is described by the polarisability derivative, $\partial \alpha/\partial Q$, where Q is the normal coordinate of the vibration. The polarisability is typically different if the field is applied parallel or perpendicular to the molecular axis or in different directions relative to the molecule. Nevertheless, for a small electric field, the polarisability is the same for the field oriented in opposite directions along the same axis ($\mu(-E) = -\mu(E)$).

Thus, the distortion induced in a molecule by an applied electric field depends on the relative orientation and returns to its initial value after a rotation of 180°. The selection rules for Raman-active vibrations are linked to molecular symmetry and identify vibrations that change a molecule's polarisability. For scattering on anisotropic crystals, with the propagating phonons dependent on the crystal axis, a similar dependence to the orientation of the light to the crystal axis is distinguishable. Raman peaks were observable within the spectra, depending on the scattering geometry.

The features of one phonon contributing to Raman scattering, incipient next to the Rayleigh peak, were followed by higher order scattered peaks in the spectra, involving multiple phonons (in the second-ordered peaks, two phonons participate). The peak intensity decreases with increasing number of participating phonons.

The population of present elementary oscillations gives the efficiency of the anti-Stokes process, proportional to the Boltzmann distribution. The stronger of the two processes is by far Stokes scattering because the population state is dominated by its vibrational state at room temperature. Most Raman techniques investigate the Stokes photons only, while ignoring the anti-Stokes photons.

2.3 Selection rules and Raman line determination

Selection rules allow one to predict if a certain molecular or lattice vibration oscillation is Raman or IR active. The selection rules for Raman scattering are analogous to those for infrared-active vibrations, which state that there must be a change in the permanent dipole moment μ during the vibration. In contrast to IR spectroscopy, the Raman active normal vibrations change the polarisability α. Group theory methods, called *nuclear site group analysis*, show that for a molecular symmetry centre, the Raman-active vibrations will be silent in the infrared and vice versa. [11, 12, 14, 15] Because of the requirement of a conserved total angular momentum in the ground state of the system, only certain transitions can be induced. A basic summary and tables are given in the book by W. J. Miller. [15]

The IR intensity is proportional to the square of the variation of the dipole moment μ, whereas the Raman intensity is proportional to the square of the change in the polarisability α:

$$I_{IR} \propto \left(\frac{\partial \mu}{\partial q}\right)_0^2, \qquad I_{Raman} \propto \left(\frac{\partial \alpha}{\partial q}\right)_0^2. \qquad (2.3)$$

Here, q is the normal coordinate of the vibration.

In crystals, the interaction between the incident photon and a phonon is indirect. The interaction is connected to electronic interband transitions, which determine the dielectric susceptibility χ in the visible spectral range. [16] If these interband transitions are influenced by phonons, Raman scattering occurs. The normal coordinate of a phonon can be expressed classically as

$$Q_j = A_j e^{[\pm i(q_j \cdot r - Q_j \cdot t)]}. \qquad (2.4)$$

The vibrational influence on the susceptibility tensor can be described by expanding χ into a Taylor series with respect to the normal coordinate of the vibration Q_j:

$$\chi = \chi^{(0)} + \sum_j \chi^{(1)} Q_j + ... = \chi_0 + \sum_j \left(\frac{\partial \chi}{\partial Q_j}\right)_0 Q_j + ... \qquad (2.5)$$

The summation runs over the $3N - 3$ vibrational modes, with N as the number of atoms in the unit cell. [16] For example, for the diatomic base of silicon, the summation runs over the three optical phonons; for SiC, the summation depends on the individual polytype.

If monochromatic light with the frequency ω_i illuminates a sample in the direction k_i, the associated electric field E induces an electric moment P according to

$$P = \varepsilon_0 \chi E = \varepsilon_0 \chi E_0 e^{[i(k_i \cdot r - \omega_i t)]}. \tag{2.6}$$

Combining equation (2.4) and (2.6) results in

$$P = \varepsilon_0 \chi_0 E + \sum_j \varepsilon_0 A_j E_0 \left(\frac{\partial \chi}{\partial Q_j}\right)_0 e^{[i(k_i \pm q_j) \cdot r - i(\omega_i \pm \omega_j) t]}. \tag{2.7}$$

The first term represents Rayleigh scattering. The second term corresponds to Stokes and anti-Stokes scattering with the frequencies $\omega_i \pm \omega_j$. This second term can only be observed if there is a change in dielectric susceptibility $(\partial \chi / \partial \omega_j)_0 = \chi^{(1)} \neq 0$. Higher order terms in the Taylor expansion describe multi-phonon processes leading to higher order Raman scattering. The elements of $\chi^{(1)}$ are often referred to as components of the Raman tensor determining a normal vibration Q_j as Raman active or not. [3, 16] The exact form of the Raman tensor depends on the crystal symmetry as described by W. Hayes [17] and Rousseau et al. [18] For example, in the orthonormal coordinate system $x = [100]$, $y = [010]$ and $z = [001]$, the Raman tensors for silicon have the form: [3, 16]

$$R_x = \begin{pmatrix} 0 & 0 & 0 \\ 0 & 0 & d \\ 0 & d & 0 \end{pmatrix}, R_y = \begin{pmatrix} 0 & 0 & d \\ 0 & 0 & 0 \\ d & 0 & 0 \end{pmatrix}, R_z = \begin{pmatrix} 0 & d & 0 \\ d & 0 & 0 \\ 0 & 0 & 0 \end{pmatrix} \tag{2.8}$$

The entire Raman scattering efficiency I is then given by

$$I = C \cdot \sum_j |e_i R_j e_s|^2 \tag{2.9}$$

where e_i and e_s are the polarization vectors of the incident and scattered light, respectively, C is a constant. [3, 16]

2.4 Influence of stress and temperature on Raman spectra

Mechanical and residual stress in semiconductor materials plays an important role in the design and reliability of devices. For example, stress can increase the probability of device failure after fabrication or during operation. Devices are often made of different materials, usually with different thermal expansion coefficients. Mechanical stress can develop in thin films and substrates, especially at the material interfaces of embedded structures. [19] Bi- and uniaxial stresses inside the material may be caused for various reasons: lattice mismatch at the interface of different materials, thermal mismatch due to different thermal expansion coefficients, or volume expansion or internal stress of different material layers deposited or grown on the substrate during device processing. Stress may directly trigger the nucleation and propagation of dislocations, generated at points with very high local stress, and the formation of cracks and voids. [20] Furthermore, stress has an important influence on dopant diffusion. Stress can cause hot carrier degradation and boron segregation into areas with dislocation loops. [3, 21] Controlling residual strain by strain engineering, however, allows the manipulation of free carrier mobility in semiconductors. For example, hole mobility may be increased by 200%. [22] In general, strain engineering may enable manipulation of electronic properties at the nanoscale, leading to custom and specialised devices. [22]

For crystalline materials, Raman spectra depend on the phonon band structure of the lattice. [3, 12, 14, 16, 17] Direct measurement of the phonon band structure can be provided by Raman spectroscopy, e.g., for silicon and silicon carbide single crystals. [3, 16, 23, 24] The phonon frequencies change with external influences, such as mechanical stress and strain. The resultant displacement of atoms changes the phonon frequencies, as indicated by shifted Raman peak centres. [3, 5, 16, 25, 26] The relation between strain ε and stress σ is given by Hooke's law, $\sigma = c\varepsilon$ where the elastic modulus c, represents the stiffness of the system. The strain is proportional to the applied stress and implies that the system can be treated as a harmonic oscillator. [26] The effect of strain on the Raman-active optical phonons of crystalline materials has been studied quite extensively since the 1970s. For a given stress, the difference ω_j between each Raman frequency ω_0 can be calculated from the eigenvalues λ_j

$$\lambda_j = \omega_j^2 - \omega_{j0}^2 \quad \text{or} \quad \Delta\omega_j = \omega_j - \omega_{j0} \approx \frac{\lambda_j}{2\omega_{j0}} \qquad (2.10)$$

with $j = 1, 2, 3$ for the given stress and $j = 0$ without stress. [3] Each of the split phonons exhibits its own frequency, which varies linearly with the components in the elastic regime. Observation of a shift may therefore be attributed to a tensile or compressive strain present in the scattering volume for moderate changes. In summary, the relation between the measured Raman shift and uniaxial stress is then obtained to be

$$\Delta\omega = \frac{1}{\omega_0} \left(pS_{12} + q(S_{11} + S_{12}) \right) \sigma_0. \qquad (2.11)$$

The elastic constant S_{ij} is a matrix element of the elastic compliance tensor S, and p and q come from the phonon deformation potentials (PDPs), all of which are material properties. [16, 26] Optical phonons are influenced by strain in the material via the phonon deformation potentials. The amount of stress σ is determined via the elasticity constants from external or internal stress according to Hooke's law $\epsilon_{ij} = S_{ij}\sigma$ (where ε_{ij} is the strain tensor component). Compressive stress ($\sigma < 0$) leads to a positive shift ($\Delta\omega > 0$), whereas tensile stress ($\sigma > 0$) shifts to lower wavenumbers ($\Delta\omega < 0$).

The theoretical background of the influence of mechanical stress on optical phonon modes in silicon is described in detail in the dissertation of Dombrowski [16] and the review of De Wolf. [3] The morphic effects of stress on the optical phonon modes of silicon were addressed by Ganesan [27] and Anastassakis [28]. In the presence of a symmetric strain, the triple degeneracy of the $k = 0$ optical silicon phonon with frequency is lifted due to anisotropic changes in the lattice constants. [3] For uniaxial or biaxial stress, the shift can be obtained theoretically. [29] According to the literature, the value of the constant in Eq. 2.12 was determined to be -434 or -518. [19, 30] As result, the relation measured on a (110) silicon surface can be written as follows

$$\sigma \text{ (MPa)} \approx C \cdot \Delta\omega \text{ (cm}^{-1}) \qquad \sigma \text{ (MPa)} = -434\Delta\omega \text{ (cm}^{-1}) \text{ uniaxial} \quad (2.12)$$

$$\sigma_{xx} + \sigma_{yy} \text{ (MPa)} = -434\Delta\omega \text{ (cm}^{-1}) \text{ biaxial.}$$

Mechanical stress leads to a change in the phonon frequency but should not influence the phonon lifetime (no peak broadening). [31] Line broadening due to mechanical stress is negligible because the phonon population density remains approximately static for moderate stresses. [26, 32, 33] However, high mechanical stress can slightly influence the phonon occupation. [34–36] The line width is expected to reveal the influence of the phonon dispersion on decay processes due to the changes in the phonon frequency throughout the Brillouin zone. [26] The mechanical component of the measured peak shift correlates with the elastic deformation of the crystal lattice and is mainly independent of temperature. [32] The transverse optical phonons, in contrast, remain unchanged with increasing pressure. [34, 37–40]

The Raman peak centre (frequency) not only depends on mechanical stress (σ) or strain (ε) but also depends on the local sample temperature (T) because the interatomic distance changes due to thermal expansion. With a linear expansion of the equilibrium position of the atoms caused by thermal expansion, the effective bond strength changes, modifying the phonon frequencies. As a result, the Raman peak position is a function of strain and stress as well as temperature. The influence of temperature on the Raman spectra of wurtzite-type crystals was investigated in 1977 by Mead et al. [4] The temperature-dependent function for the phonon confinement can be given as

$$\omega(q, T) = \omega(q) + \Delta\omega_1(T) + \Delta\omega_2(T). \qquad (2.13)$$

Here, $\Delta\omega_1(T)$ corresponds to the frequency shift due to phonon decay processes of phonon-phonon coupling and $\Delta\omega_2(T)$ indicates the frequency shift due to thermal expansion of the crystal lattice. [41] For example, investigations of the thermal expansion anisotropy of SiC show nonlinear changes in thermal expansion with temperature, thereby causing variation of the lattice parameter in different directions. [42] Following general temperature behaviour, for processes in which one phonon i is destroyed and two phonons j are generated or two phonons i are destroyed and one phonon j is created, one obtains:

$$\omega_i(T) = \omega_i(T=0) - \sum_j B_{ij} N_j(T), \quad \text{with} \quad N_j = \frac{1}{\exp(\frac{\hbar \omega_j}{k_B T}) - 1} \quad (2.14)$$

with N_j as the phonon occupation of the j-th mode. The term B_{ij} contains phonon frequencies and anharmonic coefficients for mode i, mode j and $i-j$ cross terms. [43] The temperature dependence of optical phonons can be distinguished between two ways. First, the phonon frequencies become altered due to thermal expansion, resulting in what is called an 'implicit effect' ($\Delta\omega_2(T)$ in Eq. 2.13), obtained from the volume dependence of the phonon frequencies. [43, 44] Second, there are also shifts in phonon frequencies arising from higher order terms (e.g., cubic, quadratic, etc.) in the crystal potential, causing an 'explicit effect' ($\Delta\omega_1(T)$ in Eq. 2.13), associated with the anharmonic interaction of the phonons. [43, 44] Anharmonic terms contribute to quantitative and qualitative corrections to the physical properties of a crystal, even though they may be small. [44] The implicit contribution due to the change in the equilibrium interatomic spacing by means of temperature may be expressed with the Grüneisen parameter for the i-th mode and the volume thermal expansion coefficient. [44] The explicit effect, in contrast, may be evaluated for the long-wavelength phonon modes through a perturbative approach. [43, 45]

Not only are the phonon frequencies influenced by temperature, but also the Raman spectral peak shape is affected. The peak shape depends on the population of the phonon modes. An increase in temperature results in changed phonon branches, leading to additional peak broadening. [46] This temperature dependence is governed by changes in the phonon occupation given by the Boltzmann distribution. [26] The temperature dependence of the half width Γ can be obtained from the following equation:

$$\Gamma(T) = \Gamma(0) \left[1 + \frac{2}{e^x - 1} \right] \quad \text{with} \quad x = \frac{\hbar \omega_0}{2 k_B T}. \quad (2.15)$$

$\Gamma(0)$ is the half-width at zero Kelvin, ω_0 is the phonon frequency, and k_B is the Boltzmann constant. [5] The line width of phonons is finite due to the anharmonic decay of a phonon into phonons of lower energy, even in a perfect crystal at zero Kelvin. [39]

In a three-phonon process, the decay of an optical phonon with frequency Ω leads to two phonons whose frequency sums to Ω, due to energy conservation. [39] Following the quasi-momentum conservation, only phonons of opposite wave vector q are allowed as final states. [39] For a perfect crystal at zero Kelvin, the full width at half maximum (FWHM, 2Γ) of the optical phonon can be described to be leading order in the anharmonic expansion as

$$2\Gamma(\Omega) = \frac{\pi}{\hbar^2} \sum_{q,j_1,j_2} |V_3(q, j_1, j_2)|^2 \delta(\Omega - \omega_{j_1}(q) - \omega_{j_2}(-q)), \qquad (2.16)$$

where the indices j_1, j_2 run over the different phonon branches, and the anharmonic matrix element V_3 is essentially the Fourier transform of the third derivative of the total energy per unit cell volume with respect to the phonon displacements. [39]

A systematic measurement of the line width and peak shift of an optical phonon in silicon over a temperature range of 5-1400 K is shown in Balkanski et al. [47] The line width and the frequency shift exhibit a quadratic dependence on temperature at high temperatures, indicating the necessity of including higher-order terms. Temperature affected peak widths in SiC have been also discussed in the literature. [48–50] At elevated temperatures, a distortion of the crystal lattice causes a change in the occupations of the phonon bands, i.e., via a decrement of the phonon decay. [46, 51] The relative intensity of the Stokes or anti-Stokes shift of the optical phonon deviates significantly from the value involving only the phonon occupation number, in particular at high temperatures. [52] The surface temperature of a Raman spectroscopy sample influenced by laser heating should provide an unambiguous phonon decrease. [52] An asymmetric broadening of the Raman peak can occur for a non-uniform local temperature distribution. [31, 41, 53] For example, such a broadening is the case for a non-perpendicular incident laser beam, resulting in non-circular local heat heating.

For the sake of completeness, it should be mentioned that, except for the resonant Raman scattering processes, the temperature can be deduced from the ratio of Stokes and anti-Stokes shifts where intensity strongly depends on temperature. Because the Stokes and anti-Stokes shifts correspond to a generation and absorption of a phonon, respectively, the ratio of their intensities within a spectrum is given by

$$\frac{I_{AS}}{I_S} = \frac{N_0}{N_0 + 1}, \qquad \text{with} \qquad N_0 = \frac{1}{\exp(\frac{\hbar \omega_0}{k_B T}) - 1}. \qquad (2.17)$$

Influence of stress and temperature on Raman spectra

N_0 is the equilibrium occupation of the $q = 0$ optical phonon of frequency ω_0. [5] Also, it should be mentioned that the measured line width Γ is always a convolution of the natural FWHM line width Γ_0 with the intrinsic instrumentation broadening Γ_i, [4] given by

$$\Gamma = \Gamma_0 + \sqrt{\left(\frac{\Gamma_0}{2}\right)^2 + \Gamma_i^2}. \tag{2.18}$$

Peak shifts caused by temperature and mechanical effects can be separated because temperature causes a distortion of the lattice and changes the phonon population, whereas mechanical effects only alter the lattice without changing the phonon population. Depending on the direction of mechanical stress (compressive or tensile), thermally and mechanically induced peak shifts can compensate or intensify each other. The following table shows a principle cause/effect diagram, neglecting additional impacts due to possible nonlinear effects.

variation	peak centre position ω	peak width Γ
tensile stress (strain)	⇓	⇒
compressive stress (strain)	⇑	⇒
increasing temperature	⇓	⇑
decreasing temperature	⇑	⇓

Note that the precise calculation of the local strain from the shift of the phonon frequencies depends in many cases on crystal orientation. Although the excitation laser has a polarisation ratio of 100:1, all measurements discussed in the following were taken unpolarised (no analyser was used), and the inferred local strain corresponds to a scalar average rather than exact components of the strain tensor.

3 Methods

3.1 The confocal Raman microscope set up

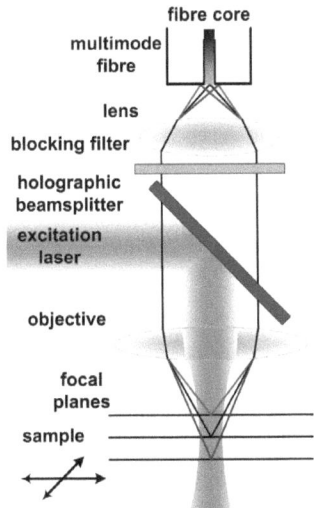

Figure 3.1: *Schematic of the confocal setup. Shown are the principal optical components and the blocking of out-of-focus photons.*

Confocal microscopy offers several advantages in comparison to conventional wide field optical microscopy. The axial focus (focal depth) can be controlled directly, thus reducing background signals from outside the confocal volume. A pinhole in the optical pathway blocks out of focus photons (Fig. 3.1). The confocal images are generated by scanning the sample and recording a full Raman spectrum at each image point. A commercial confocal Raman microscope (alpha 300 R by WITec; Ulm, Germany; www.witec.de) was used for all measurements. With a standard 100× objective (working distance: 0.26 mm; NA = 0.90), a diffraction-limited focusing of the excitation laser (Nd:YAG, SHG 532 nm, $P_{max} = 22.5$ mW) is realised. The confocal set up in backscattering geometry was achieved using the core of a multimode fibre as a pinhole that transmits the light to the spectrometer. The fibre is mounted in the image plane of the microscope and can be laterally adjusted for maximum collection efficiency. The focal depth can be adjusted by exchanging the fibre with one of smaller or larger core diameter.

Elastically scattered photons and reflexions were blocked by a sharp edge filter. Full spectra were acquired on each image point with a lens-based spectrometer with a back-thinned CCD camera (1024 × 128 pixels, cooled to -65 °C) with a 19 ms minimum integration time. With an 1800 mm^{-1} grating, the spectral resolution was 1.17 cm^{-1} per CCD-pixel. Sub-pixel spectral resolution was achieved by fitting a Lorentz function to the Raman peaks. The samples were scanned with a piezo-driven scan stage (100 × 100 × 20 µm^3) below the objective.

3.1.1 Pinhole influence and light suppression

The choice of the pinhole size is crucial because the signal should be intense, implying a large diameter, but the image resolution should be as high as possible, implying a smaller diameter.

The response of an imaging system to a point source is given with the point spread function (PSF). Here, the PSF is mainly determined by the microscope objective. The effective PSF for the detection is a convolution of the optical PSF (coming from the optical components) with the detector function of a real two-dimensional detector with radius v_P. Figure 3.2 shows the lateral resolution of a scattering point v_{FWHM} as a function of detector size v_P.

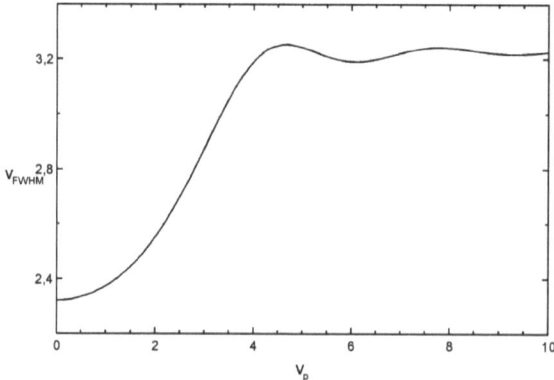

Figure 3.2: *Lateral resolution of a scattering point v_{FWHM} as a function of detector size v_P, taken from [54], reproduced with permission.*

The size of the pinhole, in optical coordinates, should not exceed $v_{P_{max}} = 2.5$ to avoid a loss of depth resolution. To obtain the highest lateral resolution, the pinhole size should be below $v_{P_{max}} = 0.5$. [54] In practice, the pinhole size can be up to $v_{P_{max}} = 4$ without significantly changing depth resolution and up to $v_{P_{max}} = 2$ without significantly changing lateral resolution. The relationship listed below,

$$\frac{M}{NA} \geq \frac{\pi d_0}{v_{P_{max}} \lambda} \qquad (3.1)$$

must be satisfied, where M is the magnification, d_0 is the pinhole diameter and NA is the numerical aperture of the objective. Beam path and objective specify the left side of the equation. The parameter M/NA calculated for several objectives is shown in the following table (taken with permission from [54]).

The confocal Raman microscope set up   19

objective	10/0.25	20/0.40	40/0.6	60/0.80	100/0.90	100/1.25	100/1.40
M/NA	53	50	67	75	111	80	71

Calculating the right side of equation 3.1 using $v_{P_{max}} = 2.5$ and a wavelength of 532 nm, the value is 59 for the fibre with a core diameter of $d_0 = 25$ µm, 118 for 50 µm and 236 for 100 µm. [54] As result, for the 100×, NA = 0.90 objective, the fibre with the 50 µm core diameter has the best compromise between resolution and light efficiency.

3.1.2 Gaussian laser focusing and beam parameters

When measuring anisotropic samples such as silicon carbide, where the Raman spectra depend on the orientation of the incident laser to the crystal axis, a small amount of Raman scattering can occur with phonon modes propagating along a different axis than the chosen one. A non-zero angle between the crystal axis and the incident photons may occur due to an imperfect alignment of the sample to the laser. This systematic error can be minimised by careful adjustment. An additional source of intrinsic error may occur as well: depending on the pinhole diameter, scattered light may be recorded, which comes from phonon modes propagating in non-chosen directions.

Laser beams can often be described with a Gaussian function (fundamental or Gaussian modes with TEM_{00} intensity distributions within the beam profile). Gaussian beams are usually considered in situations where the beam divergence is relatively small, so the paraxial approximation can be applied. This approximation allows one to disregard the term with the second-order derivative in the propagation equation (as derived from Maxwell's equations), so that a first-order differential equation results. For a monochromatic beam propagating in the z direction with the wavelength λ, the complex electric field amplitude is

$$E(r,z) = E_0 \frac{w_0}{w(z)} \exp\left(-\frac{r^2}{w(z)^2}\right) \exp\left[-i\left(kz - \arctan\frac{z}{z_R} + \frac{kr^2}{2R(z)}\right)\right] \quad (3.2)$$

where $|E_0|$ is the peak amplitude, w_0 is the smallest beam radius (beam waist), $k = 2\pi/\lambda$ is the wavenumber, Z_R is the Rayleigh length, and $R(z)$ is the radius of curvature of the wavefronts. The oscillating real electric field is obtained by multiplying the phasor with $\exp(i2\pi ct/\lambda)$ and taking the real part. A plane wave front of the light can be assumed in the range of the Rayleigh length Z_R around the beam waist w_0. The Rayleigh length is defined as the distance from the beam waist in the propagation direction, where the beam radius is increased by a factor of $\sqrt{2}$. For a circular beam, the cross sectional area is doubled at this point. The field distribution around the focus of a Gaussian beam is shown in Fig. 3.3.

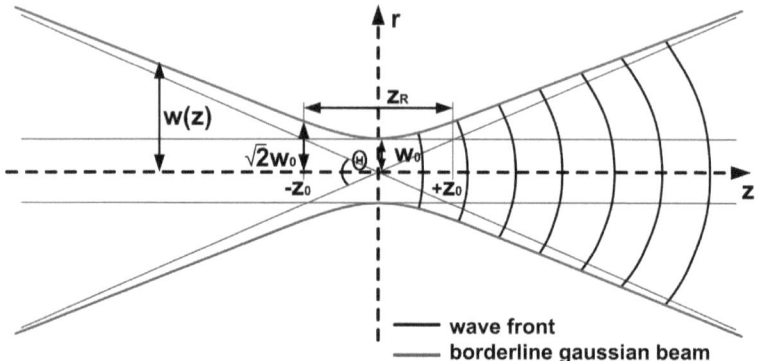

Figure 3.3: *Illustrated Gaussian beam of a focused laser beam with the parameters of beam waist w_0 and Rayleigh length Z_R.*

The Rayleigh length is determined by the waist radius w_0 and wavelength λ:

$$Z_R = \frac{\pi w_0^2}{\lambda} \qquad (3.3)$$

where the wavelength λ is the vacuum wavelength divided by the refractive index n of the material. For beams with imperfect beam quality and a given waist radius, the Rayleigh length is effectively decreased by the beam quality factor (M^2).
The beam radius varies along the propagation direction according to

$$w(z) = w_0 \sqrt{1 + \left(\frac{z}{Z_R}\right)^2}. \qquad (3.4)$$

The Rayleigh length Z_R, as show in equation 3.3, determines the length over which the beam can propagate without diverging significantly. The achievable focal spot $w(f)$, when a collimated Gaussian beam is focused by a lens with focal length f, can be calculated with the equation

$$w(f) = \frac{\lambda f}{\pi w} \qquad (3.5)$$

where it is assumed that the beam radius at the focus is much smaller than the initial beam radius w. This condition is violated for beams with an overly small incident diameter; the focus is then enlarged. [55] Also, it is assumed that the beam radius

before focusing is significantly larger than the wavelength λ, so that the paraxial approximation is valid.
Here, the commonly used fibre with the 50µm core diameter results in a confocal depth of 1 µm. With a Rayleigh length of about 0.5 µm when using the (100×, NA = 0.90) objective, additional peaks arising from other crystal orientations may be present in the spectra.

3.2 Local heating due to the focused laser beam

Using the 100× objective, the very strong focused laser reaches power densities of a few MW/cm^2 within the focal spot. This high power density may affect the specimen directly. Heating of silicon by pulsed and continuous wave (cw) lasers was demonstrated in the 1980s. [56, 57]
Light absorbed by the material can lead to an increase in the local temperature distribution, depending on the thermal conductivity. This results in a local heating effect, which can be very large at small free-standing structures or at edges. The energy absorbed in the sample material is

$$Q(r,t) = I_0(1-R)\exp\left(-\frac{r^2}{\omega_0^2}\right) \tag{3.6}$$

where R=R(T,λ) is the reflectivity, I_0 is the intensity and ω_0 is the radius of the focused laser. [29]
The recorded Raman scattering originates mostly from the boundary area of the beam profile, which does not usually correspond to the area of maximal temperature but to an area of maximal scattering efficiency. [31] Thus, the spectra represent an average over the whole laser spot.
With increasing local temperature, thermally induced mechanical stress may appear in the sample. Either an additional line shift due to the locally heated matter (in the case of silicon and silicon carbide) or, in the case of titanomagnetites, even an alteration due to heat-induced local oxidation may occur.

3.3 Image generation

Two filtering methods were used for image generation from Raman spectra. Spectra were measured on each image point, for example in a 100 × 100 point matrix. By integrating the counts in a selected range of wavenumbers around a prominent peak in each spectrum, compositional images can be created. In a more sophisticated approach, a Lorentzian peak fit is applied to each spectrum to identify changes in peak centre and peak width due to, for example, temperature variation or mechanical stress. By plotting the peak centre, the resulting image represents the stress distribution qualitatively, whereas using the relative variation of the peak width reflects the temperature distribution. Figure 3.4 shows marked regions in a typical spectrum for sum filtering (I and II) and for a Lorentz fit (III). The filters result in a compositional image of titanomagnetite with varying titanium content in a geological sample (Fig. 3.4 right), described in section 4.4.

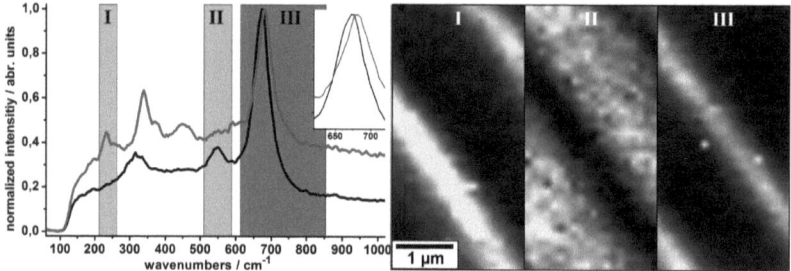

Figure 3.4: Raman spectra and compositional images of titanomagnetite with filtered regions for sum filtering (light grey, I and II) and for Lorentz fits (dark grey, III). The inset highlights the line shift between spectra. The compositional images are generated with filters I, II and III.

4 Spectroscopic characterisation

4.1 Silicon

Silicon is by far the most used and best investigated of the semiconductor materials. It has the same diamond cubic crystal structure as diamond, with each silicon atom covalently bonded in between four neighbours. Positively or negatively doped silicon (p- and n-type silicon) is used in electronic devices such as transistors for integrated circuits (ICs), flash memory and microprocessors. Because the properties and manufacturing methods of silicon are well known, silicon is used not only in electronic circuits but also in micro-electro-mechanical systems (MEMS) for sensor technology applications as a cheap and easy-to-use substrate material.

Raman spectroscopy can provide information about stress and local temperatures in a device during operation and residual stress after fabrication. These parameters are important to optimise manufacturing and increase reliability of the device.

4.1.1 Phonon dispersion curve and band structure

Raman scattering in silicon occurs on phonons propagating in the crystal. Figure 4.1 shows the electronic band structure (left) and phonon dispersion curves (right) of silicon, which were calculated in 1979. [58] The conduction band minimum is at 85% of the distance between the Γ-point, the centre of the first Brillouin zone and the X-point, the intercept point of the body diagonal with the first Brillouin zone in the Δ direction. The smallest direct band gap of $E_{G_1} = 3.4\,\text{eV}$ (corresponding to a wavelength of $\sim 370\,\text{nm}$) is between the valence band Γ'_{25} and the conduction band Γ_{15}. Another direct energy level transition is located at $E_{G_2} = 4.2\,\text{eV}$ ($\sim 300\,\text{nm}$). The indirect band transition is at $E_g = 1.1\,\text{eV}$ and is important for absorbing light. Information about the band structure can be also obtained from the wavelength-dependent complex refraction index and extinction index (Fig. 4.2). The direct energy level transitions can be distinguished as an increase in the extinction coefficient. Interactions between the incident laser light and the propagating phonons in silicon during Raman scattering were mediated via the electronic system. [16]

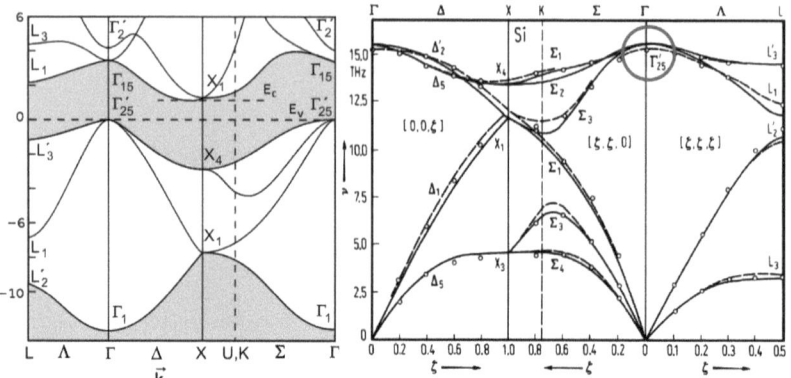

Figure 4.1: Electronic band structure (left). The unoccupied areas are in grey. Phonon dispersion curves (right) of silicon. The red circle indicates the Γ-point of the phonons on which scattering occurs, taken from [59] and [16], reproduced with permission.

The penetration depth of light in silicon depends on the wavelength of the photon, as shown in Fig. 4.2. Both the penetration depth and the absorption coefficient vary with the photon energy. At the direct band gap at 3.4 eV (∼370 nm), the absorption increases, which results in a decreased penetration depth.

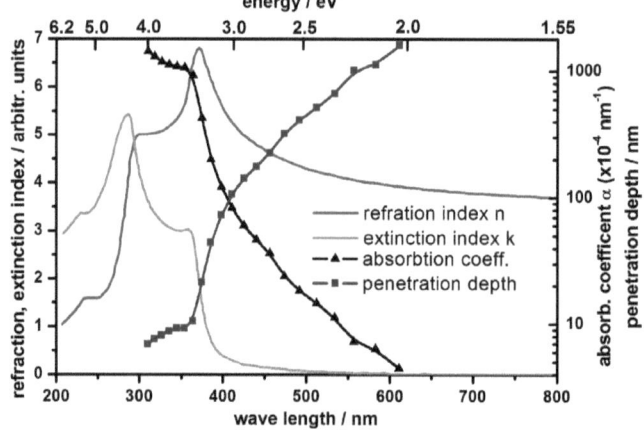

Figure 4.2: Complex refraction index (red curve) and extinction index (green curve) of silicon at room temperature. The blue curve shows the penetration depth in silicon and the black curve gives the corresponding absorption coefficient, adapted from [59] and [60], reproduced with permission. Copyright 1983 by The American Physical Society.

The penetration depth and absorption coefficient determine the local heating of the material under the focused laser. For near-surface investigations, the scattering region can be limited to the area of interest by choosing an adequate laser wavelength.

4.1.2 Raman scattering in silicon

Because of the cubic symmetry of silicon, the optical phonon branches are triply degenerate at the Γ-point, corresponding to about $521\,\text{cm}^{-1}$ at room temperature with F_{2g} symmetry (marked with a red circle in Fig. 4.1), regarding the conversion factors of

$$8.0655\,\text{cm}^{-1} \triangleq 1\,\text{meV} \triangleq 0.2418\,\text{THz} \tag{4.1}$$

where the Γ-point is at 15.5 THz. [16] The triple degeneracy of the silicon phonon can be lifted by mechanical stress. [1, 28]

Three Raman tensors exist for silicon, given as $x = [100]$, $y = [010]$ and $z = [001]$ in the crystal coordinate system (see Eq. 2.8). A phonon can be considered longitudinal or transversal depending on the surface from which the scattering is observed. R_x and R_y correspond to the transverse optical phonons (TO) polarised along x and y. R_z corresponds to the longitudinal optical phonon (LO) polarised along z in (001) backscattering, whereas R_x corresponds to the LO phonon for (100) back scattering. [3, 16] This influence of crystallographic orientation on the Raman spectra was investigated by means of polarisation measurements. [61]

The following table gives the polarisation selection rules for backscattering from a (001) and (110) silicon surface as indicated in equation 2.8 and 2.9, taken with permission from [3].

Polarization		Visible		
e_i	e_s	R_x	R_y	R_z
Backscattering from (001)				
(100)	(100)	-	-	-
(100)	(010)	-	-	×
(1-10)	(1-10)	-	-	×
(110)	(1-10)	-	-	-
Backscattering from (110)		-	-	×
(1-10)	(001)	×	×	-
(1-10)	(1-10)	-	-	×
(001)	(001)	-	-	-

Only two polarisation configurations permit Raman scattering for (001) backscattering: [3, 16]

1. e_i and e_s are perpendicular: z(x,y)-z with x parallel to (100)

2. e_i and e_s are parallel: z(x,x)-z with x parallel to (110).

In both cases, only the LO phonons can be observed. To observe the TO phonons, a different scattering geometry, such as 90° scattering, or backscattering from (110), would be needed.

4.1.3 Local heating

The absorption coefficient is approximately $9 \cdot 10^3$ cm^{-1} for silicon at 532 nm, which implies a penetration depth of approximately 1 µm (Fig. 4.2). In such a small absorbing volume, the temperature increases and may affect the spectra. To clarify the influence of heating on peak centre and peak width, a piece of silicon wafer was scanned while increasing the laser power stepwise as shown in Fig. 4.3.

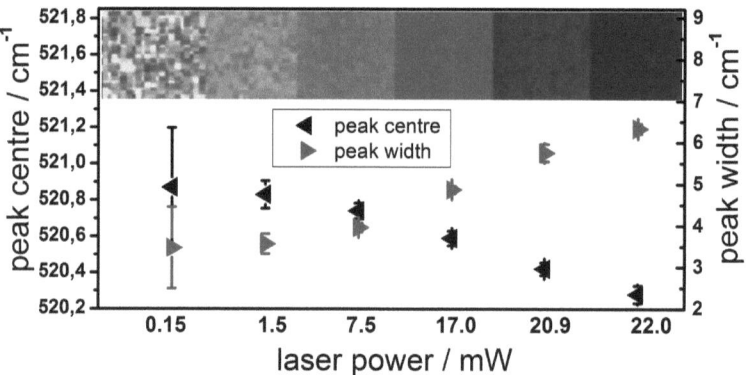

Figure 4.3: Variation of peak centre and peak width determined with a Lorentzian fit. Data were obtained in a lateral scan with stepwise increasing laser power. The fit was performed on the average spectrum for each step.

Given the possibly non-uniform local heating and spectral averaging, the peak shift alone cannot be used to estimate the maximal surface temperature because the maximum detected shift is 10 times less than expected from the centre temperature. [31] Because stress can be imposed on the crystal lattice by both mechanical loading and thermal heating, pure mechanical stress cannot be measured directly with a focused laser beam that simultaneously heats the sample. [31, 33] The inset in Fig. 4.3 shows the calculated peak centre obtained from a Lorentzian fit on each spectrum. The cross-sections show the corresponding peak centre shift and peak width. As discussed in Section 2.4, the results may be corrected from the local heating effect for a moderate laser power.

4.1.4 Experimental results

It was possible to visualise the dynamic bending stresses in a silicon micro-cantilever, which was driven at its flexural eigenmode. For the stress measurement, the data were corrected for the local heating caused by the focused laser beam. Figure 4.4 shows the uncorrected shifts with corresponding peak width. The corrected stress distribution in the cantilever with and without excited oscillation is shown below in Fig. 4.4(c).

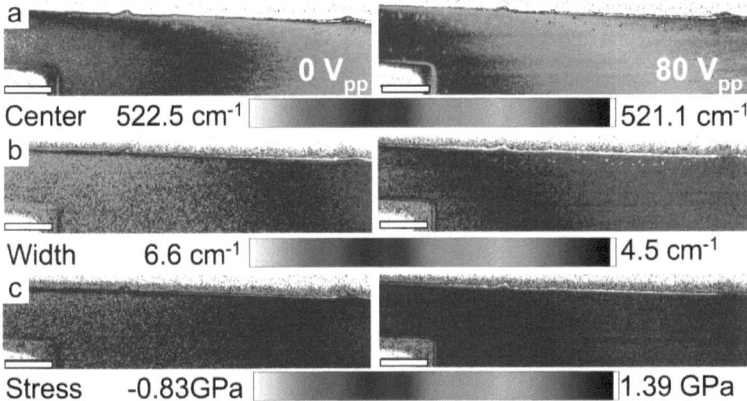

Figure 4.4: Measurements of a silicon AFM cantilever driven at its first eigenfrequency, taken from [62], reproduced with permission from Elsevier.

The results clearly indicate that the characterisation of the stress distribution in a vibrating micro structure is accessible by micro Raman spectroscopy. This demonstrates the usability of confocal Raman to visualise appearing stress fields even while a device is operating and the measurement correction of local heating.

4.2 Silicon carbide

Silicon carbide (SiC), a robust semiconductor, is used in applications with heavily demanding working conditions. SiC can be used for devices with high operating temperatures and high voltages or in optical applications into the ultraviolet region. This universal utility leads to a high interest in its physical properties. [38,63,64] Its very high thermal conductivity (~ 5.0 W/cm), high saturated electron drift velocity ($\sim 2.7 \cdot 10^7$ cm/s) and high breakdown electric field strength (~ 3 MV/cm) makes SiC a material of choice. [64,65] The strong Si-C bonding makes SiC resistant to chemical attack and radiation.

For example, SiC is used in insulated-gate bipolar transistors as a semiconductor for pulsed power applications, switching several megawatts of electrical power in several nanoseconds. [64] As a sensor material, SiC is able to work reliably in conditions where other materials may malfunction. These electronic devices will offer more efficient and cheaper systems and open a new generation of electronic equipment not requiring cooling and transformers. As with silicon, Raman spectroscopy can provide information for improved manufacturing parameters giving reduced residual stress or optimisation of the device during operation.

4.2.1 Crystal structure and polytypes

Silicon carbide forms long-range ordered polytypes that are stable or metastable at room temperature. So far, more than 250 polytypes of SiC are known that have the same chemical composition, but show a different unit cell structure, with the variation of three stacking layers, labelled A, B, and C. The unit cell structure can be cubic (C, zincblende structure), hexagonal (H, wurtzite structure) or rhombohedral (R). SiC has a tetrahedral bonding of silicon with four nearest carbon neighbours. The most basic SiC structure is the cubic 3C-SiC (also named as β-type; for historic reasons any non-cubic polytype is called α-type). The crystal polytypes are designated according to the number of double layers in the unit cell and the letter of the lattice type. Figure 4.5 shows the layer stacking for some SiC polytypes. For each, the length of the unit cell along the c-axis is n times larger than of the basic 3C-SiC, resulting in the nomenclature of nH-SiC and $3n$R-SiC polytypes. For example, the common polytype 4H-SiC (with ABCB stacking) consists of four units in the c direction with the lattice parameters a = 0.30806 nm and c = 1.0087 nm. [66]

Four hexagonal Miller indices were used, describing the directions in all SiC polytypes except for the 3C polytype, where the normal cubic notation is used. The last hexagonal index refers to the c-direction, whereas the three first describe directions in the basal plane with a 120° angle between the axes.

By observing the SiC crystal from the side, the stacking sequence can be projected as in Fig. 4.5, revealing a zig-zag pattern. However, the surrounding lattice does not look the same for each position. The A position has a different surrounding lattice than the B and C positions. The other two positions (B and C) are called cubic, k_1 and k_2. In 3C-SiC there is of course only one cubic site and in 2H-SiC only one hexagonal site. Thus, the 6H-SiC polytype can be characterised as 33% hexagonal, whereas the 4H- and 2H-SiC polytypes are 50% and 100% hexagonal, respectively. [67]

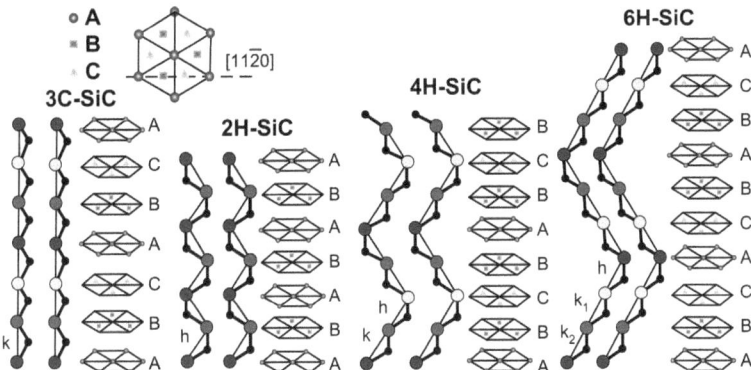

Figure 4.5: Illustration of the $[11\bar{2}0]$ plane and layer stacking (ABC) of silicon carbide for the polytypes 3C, 2H, 4H and 6H; adapted from [67, 68]] and reproduced with permission.

4.2.2 Phonon structure and mode folding

Group theory analysis shows that the Raman active modes of the basic structure are the A_1, E_1 and E_2 modes. [2, 17] Phonons caused by atomic motion parallel to the c-axis are called axial (longitudinal, L), perpendicular to the c-axis planar (transversal, T) having an acoustic (A) or optical (O) origin.

For polytypes of higher order than 3C-SiC, the increased unit cell size results in a decreased Brillouin zone Γ ($q = 0$) in the Γ-L direction to $1/n$ of its basic size with a reduced wave vector ($x = q/q_B$) of the phonon mode, as shown in Fig. 4.6. The dispersion curves of the propagating phonons are folded back into the basic zone, illustrated in Fig. 4.7. [2, 69] These folded curves create new phonon modes at the Γ-point and emerge corresponding to the phonons inside or at the edge of the basic Brillouin zone. These additional modes are referred to as "folded modes", labelled with "F". The symmetry of these peaks can also be assigned by group theory. [69] The lattice dynamics and phonon dispersion curves of SiC were modelled and compared with experimental data, resulting in an analytical model (bond-charge model) [70] supplemented with thermal processes like thermal vibrations and the displacement of atoms in the lattice.

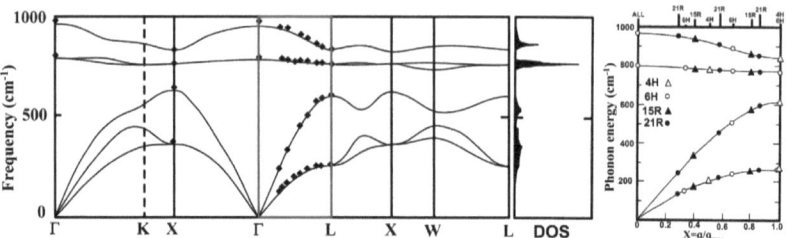

Figure 4.6: *Phonon dispersion curve of 3C-SiC (left). The Γ-L region is marked in red with corresponding density of states (DOS). Right: combined dispersion curves for different polytypes, reprinted with permission from [23, 70]. Copyright 1968, 1994 by The American Physical Society.*

The A_1 and E_1 phonons are split into longitudinal optical (LO) and transverse optical (TO) modes. [69] This splitting originates from an additional force on the permanent dipole moment, for the longitudinal displacement only. Folded optical modes exhibited less splitting than acoustic modes. [2, 23] The polarisation and geometry of the laser light to the c-axis (back- or right-angle scattering, for the a-axis swapped direction) change the measured Raman spectra, showing varying features, e.g., the $E_1(LO)$ line is only present in one direction. [2, 23, 49]

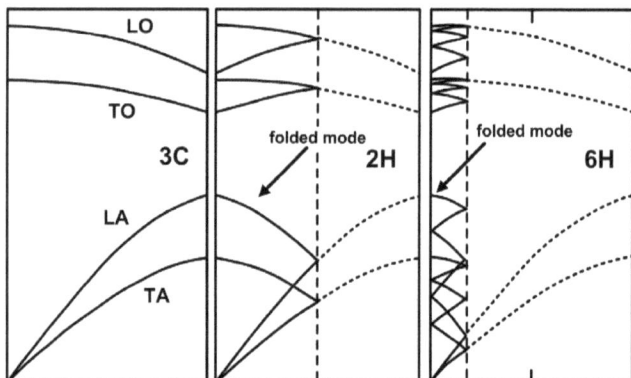

Figure 4.7: *Mode folding for 3C, 2H and 6H SiC, taken from [2]. Copyright Wiley-VCH Verlag GmbH & Co. KGaA. Reproduced with permission.*

4.2.3 Raman line shift

The Raman spectra show different features for each polytype and depend on the orientation of the incident laser beam to the crystal axis (Fig. 4.8). Raman spectra of undoped 6H-SiC, undoped and nitrogen doped 4H-SiC polytypes with the laser beam oriented parallel and perpendicular to the c-axis are illustrated.

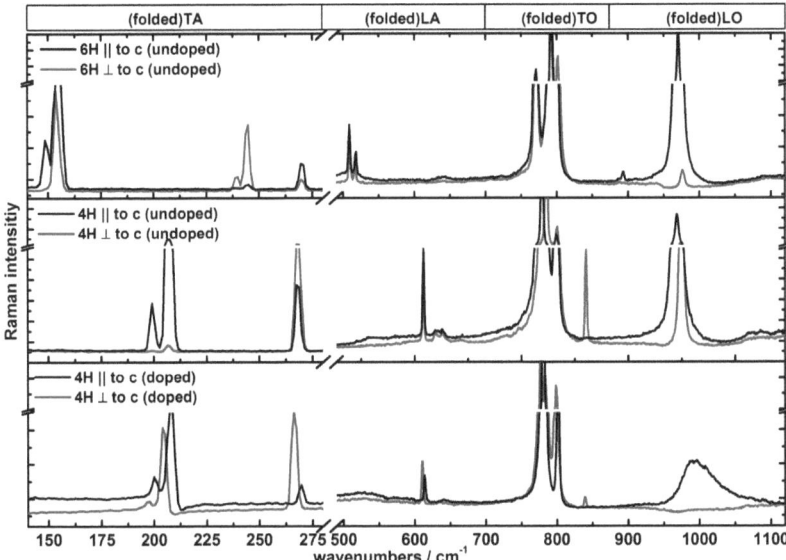

Figure 4.8: Raman spectra of undoped 6H-SiC (top), and undoped and nitrogen doped 4H-SiC (middle, bottom respectively) polytypes oriented parallel and perpendicular to the c-axis.

The thermal expansion anisotropy of SiC was discussed previously, exhibiting different values for the c and a axes and a nonlinear expansion coefficient with increasing temperature, varying multiple lattice parameters and unit cell vectors. [42, 71] The Raman peak shift with varying temperature can be used for temperature measurements. The temperature dependent Raman peak frequency can be modelled with the following equation (taken from [72]):

$$\omega(T) = \omega_0 - C\left(1 + \frac{2}{e^x - 1}\right) - D\left(1 + \frac{3}{e^y - 1} + \frac{3}{(e^y - 1)^2}\right) \quad (4.2)$$

with

$$x = \frac{\hbar\omega_0}{2k_B T}, \qquad y = \frac{\hbar\omega_0}{3k_B T}. \quad (4.3)$$

The constants C and D are fitting parameters. In $\hbar\omega_0$, the anharmonic interaction is neglected, and $k_B T$ is the thermal energy. The term starting with D can be neglected for moderate temperatures (less than 500 °C), resulting in a linear dependency of frequency on temperature. For temperatures up to 1000 °C, four anharmonic terms must be considered in the Hamiltonian model and an additional peak shift must be taken in account, as well stresses inside the bulk material. [33] The shift of the E_2(FTO) Raman peak of undoped 6H-SiC with increasing temperature was used for temperature measurements inside a light emitting diode made from SiC. [72]

The dependence of the Raman spectra of SiC on isotropic pressure has been investigated experimentally and theoretically. [35, 36, 40, 42, 66, 71, 73] It is possible with *ab initio* and group theory methods to calculate this dependence. [69, 70, 74]

With nitrogen doping as an electron donor, the spectral shape and width as well as the peak position of the A_1 phonon (LO) vary with concentration. This effect can be assigned to a plasmon-phonon coupling. [69, 74] The plasmon-coupled LO phonon (LOPC) was used to characterise ion-implanted SiC for different annealing temperatures. [75] Harima *et al.* investigated the plasmon frequency, carrier and phonon damping deduced from line shape by means of Raman scattering from anisotropic LO phonon plasmon coupling in n-type 4H- and 6H-SiC. [49] Owing to the major effect of the LO damping, the LOPC line shape can be used to obtain the carrier density and mobility. [49] Previous experimental and theoretical studies describe the isotropic pressure dependence shift of the Raman lines in SiC. Carrier mobility and concentration deduced from Raman spectroscopy show a reasonable agreement with those found by electrical measurements. [49]

4.2.4 Local heating

In contrast to silicon, the local heating effect in silicon carbide can be neglected due to the minimal influence of the laser. The change in peak centre position is in the same range for the investigated polytypes 6H-SiC (LO peak, $0.005\,\text{cm}^{-1}/\text{mW}$) and 4H-SiC (LO peak, $0.006\,\text{cm}^{-1}/\text{mW}$). For comparison, the shift in silicon is $0.03\,\text{cm}^{-1}/\text{mW}$, more than five times that of the shift in SiC. Figure 4.9 shows the peak shift and increasing peak width of silicon in comparison with the silicon carbide polytypes due to increasing laser power, from 0.1 mW to a maximum of 22.2 mW.

Silicon carbide

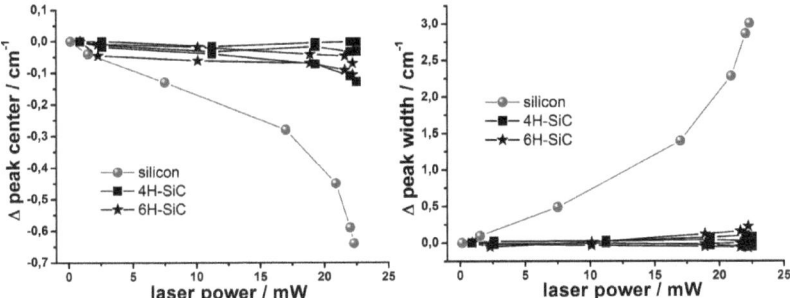

Figure 4.9: *Temperature-induced peak shift and increasing peak width of silicon in comparison to SiC polytypes. The influence of local laser heating for SiC is much smaller than for silicon.*

4.2.5 Experimental results

The peak positions of the Raman lines in silicon carbide vary linearly with temperature. The slope of the temperature shift increases with increased energy transfer due to inelastic scattering, showing a linear dependence with temperature in the temperature range of about $\Delta T = 110\,\text{K}$. Figure 4.10 left, shows the slope of the temperature shift for all Raman peaks and orientations of the investigated samples, which were used for temperature calibration purposes. Although the peak centre position is typically shifted to lower wavenumbers with increasing temperature, the nitrogen-doped plasmon-coupled longitudinal optical phonon mode (LOPC) moved to higher values. For accurate stress and temperature measurements in silicon carbide devices, the linearity of the temperature shift within the typical operation temperatures simplifies spectroscopic temperature measurements

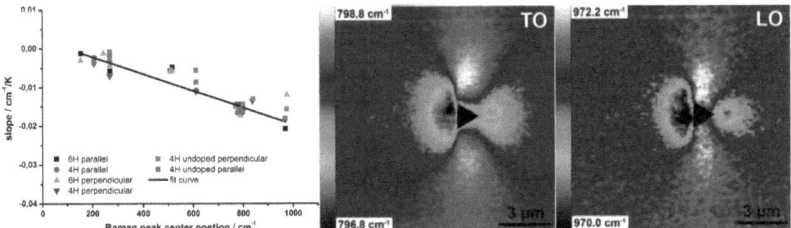

Figure 4.10: *Temperature-dependent shifting of peak centres (left) and stress fields around a nanoindentation in (0001) 6H-SiC, (right), taken from [76, 77]. Reproduced with permission, ©2009 Optical Society of America, Inc.*

The stress fields in silicon carbide caused by nanoindentations show the stress distribution in the material (Fig. 4.10, right). A comparison with s-SNOM (scattering scanning near-field optical microscope) measurements reveal a similar qualitative contrast of the strain fields around nanoindentation sites, corresponding in their spatial distribution and their qualitative sign (compressive/tensile). The Raman frequency shifts can be directly related to local strain, in much the same way as was observed with s-SNOM. The magnitude of the phonon line shift agrees well with the shift of the phonon-polariton near-field resonance. Thus, s-SNOM allows for the resolution of the smallest features of the indent down to nanocracks, whereas diffraction limited confocal Raman spectroscopy gives full spectral information for the changes in phonon dispersion. Scattering-SNOM is sensitive to the strain distribution near the surface, whereas far-field Raman scattering also occurs beneath the surface and can reveal changes in sub-surface strain distributions.

4.3 Alteration of calcium fluoride caused by UV-light

Intense ultraviolet laser radiation may cause surface damage in optical components. [78] Depending on the substrate material and coatings, optical components in a laser will start to degrade at a certain power level. [6, 79–84] The damage first appears at the exterior surface before intracavity damage occurs. The effect is proportional to the square of the field and disappears if there is a perfect antireflection coating on the surface, as first mentioned by Boling in 1973. [6] Working with deep UV light (DUV) limits the number of substrate materials that are transmissive in this spectral range. The most common UV lasers are excimer lasers, which are very intense ultraviolet light sources, generating laser pulses in the nanosecond range with high power densities. Common lasers are krypton fluoride (KrF, 248 nm) and argon fluoride (ArF, 193 nm), used for optical lithography, medical and industrial applications and in metrology. The active media are the so-called excimer (ex̲cited di̲mer) molecules, which represent a bound state of their noble gas and halogen constituents only in the excited electronic state, but not in the ground state.

The preferred material for excimer laser optical components is metal-fluoride crystals, due to their excellent optical properties, large band gap energies and resistance to the fluorine-containing laser gas. High reflective and outcoupling resonator mirrors are usually dielectric mirrors. The reflective coatings are made of an alternating sequence of $\lambda/4$ metal-fluoride layers with high and low refractive indices, commonly deposited on the internal face of the resonator. Figure 4.11 outline suitable materials for UV applications.

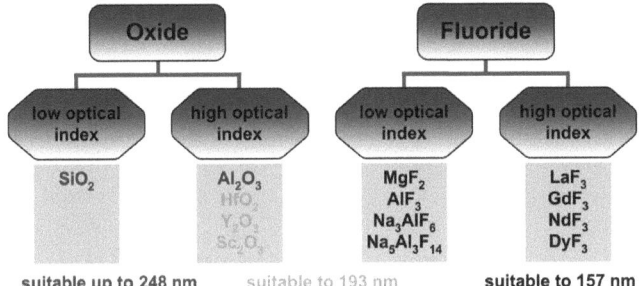

Figure 4.11: *Metal-fluoride and metal-oxide materials used in ultraviolet optical components.*

Oxide materials are not a good choice because they either absorb at the wavelength of interest or have low band gap energies and are more vulnerable to colour centre formation. However, specialised oxide materials such as high purity fused silicon dioxide (fused silica) have been shown to be useful at wavelengths below 200 nm. The only two materials that offer a desirable combination of optical, mechanical and chemical characteristics are UV grade fused silica, which transmits down to about 190 nm but is attacked by the fluorine in an excimer laser gas, and calcium fluoride (CaF_2), which can be used at wavelengths as short as 130 nm. Magnesium fluoride (MgF_2) has a similar transmission window but is transparent deeper into the UV and is harder (and thus more mechanically durable) than CaF_2. In contrast, MgF_2 is more birefringent, with a higher tendency for colour centre formation, which significantly limits its utility for transmissive optics. Nevertheless, magnesium fluoride can be used as a thin film coating in dielectric mirrors or as a protective coating.

The high power density and narrow wavelength range of lasers create problems not found in other optical systems. Nowadays, calcium fluoride is considered the preferred material for transmitting optical components in ArF laser systems. CaF_2 is available in different quality levels (IR, UV, VUV-laser grade). Excimer grade CaF_2 provides the combination of deep ultraviolet transmission, high damage threshold, resistance to colour centre formation, low fluorescence, high homogeneity, and low stress-birefringence characteristics required for the most demanding deep ultraviolet applications and therefore the best long term stability of the optical components.

Even though the material can be produced in a high purity grade and polished to surface roughness values well below $0.5\,\text{nm}_{rms}$, degradation of CaF_2 optical elements is often observed for laser operation conditions with high energy doses and pulse repetition rates. In most cases, deterioration effects start after irradiation with a certain number of laser pulses, leading to a continuous increase of optical losses.

In many applications using excimer lasers, a specific degradation effect is observed for the CaF_2 outcoupling windows, which starts at the rear surface and results in characteristic damage morphology. Until now, this damage has not been investigated in detail. Confocal Raman spectroscopy can provide spatially resolved information about the chemical composition of the exterior resonator surface. The high reflective (HR) mirror, however, shows a significantly longer optical lifetime of several billion pulses.

4.3.1 Material properties of calcium fluoride

Exposure to 100% relative humidity at room temperature does not fog polished surfaces even after one month. In normal working conditions, polished surfaces will not degrade. In dry environments, calcium fluoride can be used up to 1000 °C. In the presence of moisture, calcium fluoride surfaces will degrade for temperatures exceeding 600 °C. Calcium fluoride is inert to organic chemicals and many acids (including HF), but it will slowly dissolve in nitric acid.

A list of the CaF_2 material properties is shown in the following table; taken with permission from [85].

transmission range	0.13 - 12 µm
refractive index	1.501830 @0.193 µm; 1.40568 @4.4 µm
reflective loss	5,6% @4 µm
Reststrahlen	35 µm
density	3.18 g/cm^3
melting point	1418 °C
molecular weight	78.08
thermal conductivity	9.71 W/(mK)
specific heat	854 J/(kgK)
thermal expansion	18.87 - 20.16 ($\cdot 10^{-6}$ /°C (25°C))
Knoop hardness	158.3
Moh's hardness	4
Young's modulus	75.8 GPa
shear modulus	33.77 GPa
bulk modulus	82.71 GPa
rupture modulus	36.5 MPa
elastic coefficient	$C_{11} = 164$ GPa; $C_{12} = 53$ GPa; $C_{44} = 33.7$ GPa
dielectric constant	6.76 @27 °C; f = 100 kHz
solubility in water	0.0017 g/100g @20 °C
cleavage planes	{111}
crystal structure	cubic, CaF_2 type structure, fluoride or fluorite type structure, Fm3m O_h^5, $a_0 = 5.4623$

Alteration of calcium fluoride caused by UV-light

Even in well purged environments, however, it has been shown that metal-fluoride optical coatings and uncoated metal-fluoride surfaces degrade by reactions with atmospheric moisture and carbon dioxide. [86, 87] Current research has confirmed that uncoated CaF_2 surfaces degrade after only a few million pulses when irradiated with pulse energies above $\sim 40\,\text{mJ/cm}^2$ using 193 nm laser light. [88] Excimer lasers typically do not have a Gaussian single mode TEM_{00} beam profile, showing local non-uniformities that are 2 - 3 times the average value, thus exceeding the $\sim 40\,\text{mJ/cm}^2$ threshold for damage formation easily. As a result, either excimer lasers must operate below their maximum power levels or shorter optical component lifetimes must be accepted if the lasers are operated at higher power levels.

4.3.2 Identification of kerogeneous carbon

Raman imaging of damaged optical structures revealed the spectroscopic fingerprint of kerogeneous carbon. Raman spectroscopy has been used since the early 1970s to study carbon materials, analysing their chemical and structural composition. [89–93] A quantification of the graphitisation of organic matter (kerogeneous carbon) with Raman spectroscopy was done by Beyssac. [91] The Raman spectrum of kerogeneous carbon is split into two characteristic regions consisting of first-order Raman peaks $(1100 - 1800\,\text{cm}^{-1})$ and second-order Raman peaks $(2200 - 3400\,\text{cm}^{-1})$, as shown in Fig. 4.12.

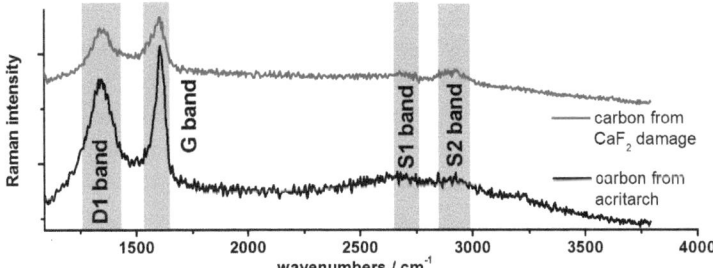

Figure 4.12: Typical Raman spectra with first- and second-order regions of kerogeneous carbon showing the most specific features (D1, G, S1 and S2 band). The upper spectrum was taken on a damaged UV optical component and, the lower on a fossilized algae.

The first Raman peak corresponds to a relative vibration perpendicular to the aromatic layers and appears at very low frequencies (about $42\,\text{cm}^{-1}$, E_{2g1} mode) and can hardly be distinguished from the Rayleigh peak. [92] The second mode (E_{2g2}) corresponds to a stretching in the aromatic layer with high frequency (about $1580\,\text{cm}^{-1}$)

normally labelled the G-band. [92] In perfect crystals such as graphite, only the G-band will appear in the first order region. For poorly organised carbon, additional bands appear around $1150\,\mathrm{cm}^{-1}$, $1350\,\mathrm{cm}^{-1}$ (D1 band and a defect band with A_{1g} symmetry from in-plane defects and heteroatoms, [94]), $1500\,\mathrm{cm}^{-1}$ (D3 band) and $1620\,\mathrm{cm}^{-1}$ (D2 band). [92] The most important features are the D1 and G bands, with the D2 and D3 band appearing as shoulders in the G band. The D2 and D1 bands are always present together and their intensities decrease with an increasing degree of organisation. The peaks in the second-order region come from overtones or combination scattering appearing near $2400\,\mathrm{cm}^{-1}$, $2700\,\mathrm{cm}^{-1}$ (S1 band), $2900\,\mathrm{cm}^{-1}$ (S2 band) and $3300\,\mathrm{cm}^{-1}$. [90, 92]

The degree of organization can be determined by the peak position, peak width and the D1/G intensity ratio (R1 ratio) with the D1/(G+D1+D2) area ratio (R2 ratio) used for describing the carbonisation process. [90–92] Present temperatures and pressure levels can be derived by analysing the R1 and R2 ratios of the kerogen Raman spectra.

4.3.3 Experimental results

UV irradiation with high energy densities of $80\,\mathrm{mJ/cm^2}$ and $193\,\mathrm{nm}$ wavelength causes surface damage within several hundred million pulses. Portions of the mirror's exterior surface are altered from the CaF_2 bulk material into $CaCO_3$ (calcite). The surface roughens and increases scattering losses in addition to the absorption losses of the altered material. The deterioration is driven by temperature and the photo-induced chemistry caused by the UV laser light. The alteration starts at the edge of the beam profile where the highest temperature gradients occur. The $CaCO_3$ crystals, which are a few micrometers in size, are embedded in a variety of organic material. Highly energetic photons induce photochemical reactions of imperfections located at the interface between the substrate and coating or with contaminants from the environment. Figure 4.13 shows the surface damage of a mirror irradiated with 250 million pulses (a) and (d) and 100 million pulses (b), (c) and (e), respectively. The lateral scan in Fig. 4.13(a) outlines the CaF_2 intensity, and (d) shows the $CaCO_3$ intensity in yellow and organics in bright blue at the same position. Figure 4.13(b) illustrate $CaCO_3$ in red and organic matter in green, and (c) kerogeneous carbon in yellow. The vertical slice in Fig. 4.13(e) illustrates CaF_2 in greyscale, $CaCO_3$ in red and organic matter in green.

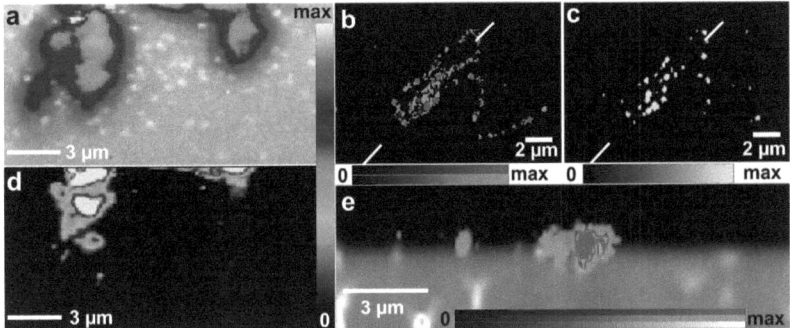

Figure 4.13: Damage structure due to UV photon irradiation, taken from [95, 96]; reproduced with permission, ©2009 Optical Society of America, Inc.

Further irradiation results in increasing absorption and, consequently, in strong thermal stress. Thus, cracks and cleavage occur at the crystallographic planes of the substrate. In summary, the failure of the resonator optics can be attributed to the influence of the flux of high energy photons on defects located at the interface between the substrate and the coating. A simple model was developed considering the initial contaminations, the emerging thermal gradient and the further migration of cracks and ruptures in conjunction with photochemical reactions. [97] The significantly longer lifetime of the HR mirror is only explainable by the assumption of equilibrium between contamination and desorption of chemical compounds. With respect to the durability of the HR mirror, the interface between substrate and coated layers seems to be the most critical part of the component.

4.4 Titanomagnetites

The alteration of titanomagnetites due to laser irradiation is another example of laser light-induced material degradation. The scientific definition of a titanomagnetite is a titanium-rich magnetite, in which Ti^{4+} replaces some of the Fe^{2+} ions. [98] Magnetite, as the base material for titanomagnetites, is a special case of ferrite with the general formula $Fe^{2+}Fe^{3+}_2O_4$ corresponding to $FeO \cdot Fe_2O_3$ (simplified Fe_3O_4) with an inverse spinel structure. [99] Titanomagnetite (Fe(Fe,Ti)$_2$O$_4$) occurs strictly as a microlitic phase. Magnetite is strongly ferromagnetic, and natural magnetite usually shows intergrowth with ulvöspinel (Fe$_2$TiO$_4$, titanomagnetite with up to 6% TiO$_2$) and ilmenite (FeTiO$_3$). The mineral structure may have some substitutional defects, where atoms in the lattice are replaced with some other atomic species. [100] The synthesis of stoichiometric synthetic titanomagnetites (Fe$_{3-x}$Ti$_x$O$_4$) is demonstrated, ranging from $x = 0$ (TM00; magnetite) to $x = 0.72$ (TM72) with a novel kind of magnetic interaction. [101, 102] For example, TM60 represents the idealised titanomagnetite with the composition $Fe_{2.4}Ti_{0.6}O_4$. Raman spectroscopy can help to quantify inhomogeneous samples by lateral compositional mapping and to classify individual minerals or varying titanium content. [103]

The identification of titanomagnetite can provide information on the conditions of rock genesis and alteration. [100] The magnetic properties of titanomagnetite, which are the principal carriers of magnetic remanence in nature, increase systematically with pressure. [104] The titanomagnetite series serves as an important carrier of the paleomagnetic record. [104, 105] Commonly, Curie points are used to determine the composition of these magnetic phases. In multi-domain titanomagnetite, the saturation of the remnant magnetisation increases as a function of titanium concentration and pressure. [106] Above about 20% titanium (TM20, $Fe_{2.8}Ti_{0.2}O_4$), magnetic coercivity also rises steeply with titanium content and pressure. Hence, titanomagnetites can be a useful magnetic thermobarometer in terrestrial investigations. [104] A discrepancy between the theoretical model and experimental data of the cation and vacancy distributions in TM60 and of the magnetic and electrical properties was shown in 1994. [107] A Raman analysis of TM60 gives some possible explanations of these differences. [108]

With Raman spectroscopy, the titanium content in mineral samples can be identified. Normalised Raman spectra of synthetic magnetite and titanomagnetites with increasing titanium content are shown in Fig. 4.14.

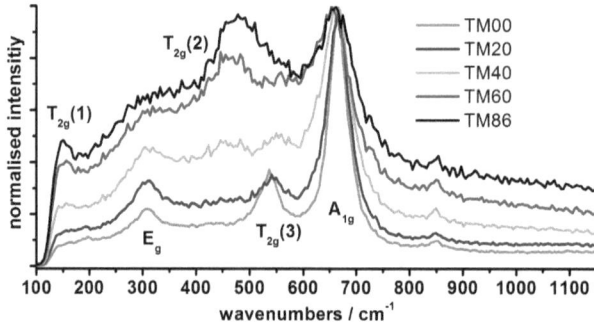

Figure 4.14: *Normalised Raman spectra taken on synthetic magnetite (TM00) and titanomagnetites with different titanium contents.*

The following table lists the measured Raman peaks of synthetic magnetite and titanomagnetites obtained by a Lorentzian fit, compared with reported peaks from different studies. [108]

Ti/(Fe+Ti)	Raman peak cm^{-1}				
Mode	$T_{2g}(1)$	E_g	$T_{2g}(2)$	$T_{2g}(3)$	A_{1g}
TM00 (magnetite) [108]	192	306	-	538	668
TM00 (magnetite)	197.8	304.4	-	536.8	664.9
TM20	193.2	309.0	-	540.7	664.1
TM40	155.8	309.2	455.1	549.7	658.7
TM60	155.7	318.2	464.2	-	647.5
TM86	154.4	-	477.7	-	654.3

Measuring titanomagnetites ($Fe_{3-x}Ti_xO_4$) with Raman spectroscopy is complicated due to the strong influence of local heating. [7] For the measurements, the laser power must be carefully chosen. For a laser-induced oxidation of magnetite to haematite, a temperature of about 240 °C with a temperature-induced peak shift was reported. [7] A series of Raman spectra, taken on natural samples and synthetic titanomagnetites, beginning with x = 0 titanium content (TM00, magnetite, Fe_3O_4) to 86% titanium (TM86, $Fe_{2.14}Ti_{0.86}O_4$) shows the incipient oxidation when a certain laser power is reached. The synthetic titanomagnetite samples were produced using the Wanamaker's floating zone technique. [107] The titanium concentration was verified by thermomagnetic analysis. [104] An oxidation of magnetite to haematite by means of thermal effects from the laser beam was reported by Shebanova et al. [7] With increasing titanium concentration, the laser power density threshold decreases (Fig. 4.15(a)). Figure 4.15(b) shows the series of the natural sample.

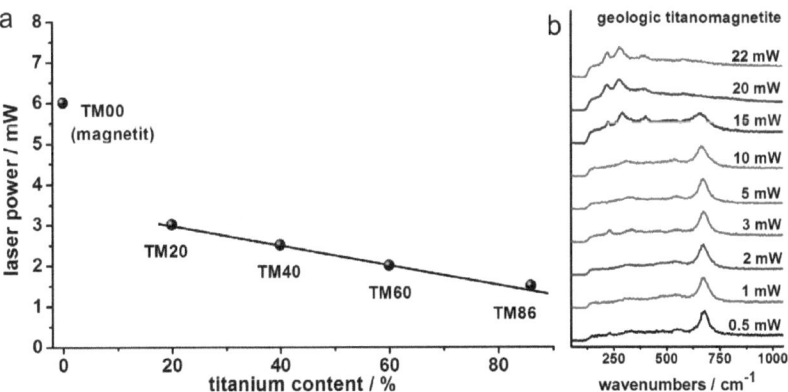

Figure 4.15: Plot of titanium concentration against laser power threshold where material alteration begins.

Due to the non-uniform local heating, the Raman spectra represent the features of both the oxidised centre and the unaffected surrounding material. Raman scattering, however, occurs in the entire area and contains the oxidised centre and the unchanged surrounding region. As a result, the measured Raman spectrum is a blend of the altered and the unaltered titanomagnetites. This is observable in Fig. 4.15(b) for a laser power of 15 mW, corresponding to about $1.1\,\mathrm{MW/cm^2}$ (100× objective).

Natural titanomagnetites from a basalt sea bed and synthetic generated titanomagnetites with varying titanium content ranging from 0 (magnetite) to 86% were investigated. In the geologic samples, alternating layers with different titanium concentrations could be identified on the scale of some micrometers for the individual layers (Fig. 4.16(d)). A change in the direction of these stripes could be found on a crack in the mineral (Fig. 4.16(c)). The optical image is shown in Figure 4.16(a) and Raman spectra from the individual stripes are shown in 4.16(b). Magnetic force microscopy measurements revealed a similar striped structure.

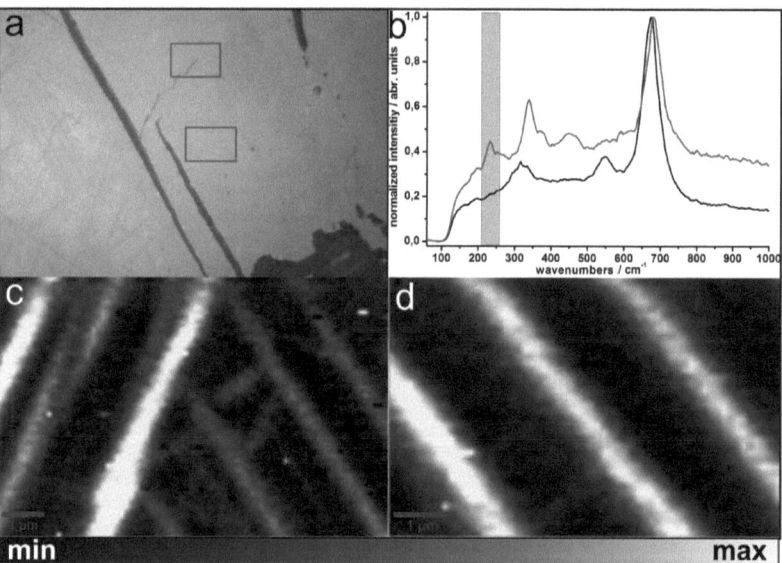

Figure 4.16: *Stripes of different titanomagnetites in a geologic sample.*

5 Summary

Confocal Raman spectroscopy is well-suited for the determination of stress fields generated in the material by means of external applied loading. By analysing the Raman peaks by means of centre position and width, additional information is accessible. For this purpose, the peaks were fitted with a Lorentzian curve. Stress causes a peak shift due to affected phonon frequencies. Through the direction of this shift, compressive and tensile stress can be distinguished. The phonon frequencies vary not only with stress, but also with changing temperatures. The population of the phonon branches is mainly influenced by temperature, which leads to an increased peak width with increased temperature. Due to the diverse impact of mechanical stress and temperature on the Raman spectra, a separation of both effects is possible.

Depending on light absorption and thermal conductivity, the sample is locally heated to some degree by the excitation laser, thus introducing additional spectral shifts, which may require compensation for more precise results. For thermal calibration purposes, spectral variations were analysed with controlled heating experiments on single crystals of silicon and different silicon carbide polytypes.

The characterisation of heterogeneous specimens was carried out on outcoupling resonator mirrors used in excimer lasers that show a UV photon-induced damage at the exterior surface when a certain energy level is exceeded. The damage structure was characterised and the components were identified, revealing an alteration of calcium fluoride bulk material to calcium carbonate (calcite). The particle size was in the range of a few micrometers in diameter with a depth of about one micrometer beneath the surface. Little filaments found on samples with the start of surface damage act as a nucleation point for further damage. The filaments were composed of organic matter and kerogeneous carbon with small embedded calcite particles.

In addition, geologic samples from a basaltic sea bed with embedded titanomagnetites were characterised for their composition. Alternating layers with different titanium ion concentrations could be identified. Synthetic titanomagnetite samples with known titanium concentrations were used as references to identify the materials. Significant local heating was observed for all magnetite and titanomagnetite samples. Above a certain threshold, laser-induced oxidation occurs, which depends on the titanium content. The laser power threshold for the oxidation decreases linearly with increasing titanium content.

In summary, confocal Raman spectroscopy has great potential to identify substances and to determine their spatial distribution within a sample. The visualisation of residual stress fields is another valuable asset. Together with the simple sample preparation and the possibility for contact-free measurement, confocal Raman spectroscopy is a valuable tool for material analysis.

6 References

[1] G. Gouadec and P. Colomban. Raman Spectroscopy of nanomaterials: How spectra relate to disorder, particle size and mechanical properties. *Progress in Crystal Growth and Characterization of Materials*, 53(1):1–56, 2007. `doi:10.1016/j.pcrysgrow.2007.01.001`.

[2] S. Nakashima and H. Harima. Raman investigation of SiC polytypes. *Physica Status Solidi a-Applied Research*, 162(1):39–64, 1997. `doi:10.1002/1521-396X(199707)162:1<39::AID-PSSA39>3.0.CO;2-L`.

[3] I. De Wolf. TOPICAL REVIEW: Micro-Raman spectroscopy to study local mechanical stress in silicon integrated circuits. *Semiconductor Science Technology*, 11:139–154, February 1996. `doi:10.1088/0268-1242/11/2/001`.

[4] D. G. Mead and G. R. Wilkinson. The temperature dependence of the raman effect in some wurtzite type crystals. *Journal of Raman Spectroscopy*, 6(3):123–129, 1977. `doi:10.1002/jrs.1250060305`.

[5] T. R. Hart, R. L. Aggarwal, and B. Lax. Temperature Dependence of Raman Scattering in Silicon. *Physical Review B*, 1(2):638–642, 1970. `doi:10.1103/PhysRevB.1.638`.

[6] N. L. Boling, M. D. Crisp, and G. Dubé. Laser Induced Surface Damage. *Appl. Opt.*, 12(4):650–660, 1973. `doi:10.1364/AO.12.000650`.

[7] O. N. Shebanova and P. Lazor. Raman study of magnetite (Fe_2O_3): laser-induced thermal effects and oxidation. *Journal of Raman Spectroscopy*, 34(11):845–852, 2003. `doi:10.1002/jrs.1056`.

[8] C. V. Raman and K. S. Krishnan. A new type of secondary radiation. *Nature*, 121:501–502, 1928. `doi:10.1038/121501c0`.

[9] G. Landsberg and L. Mandelstam. A new occurrence in the light diffusion of crystals. *Naturwissenschaften*, 16:557–558, 1928.

[10] P. Colomban. Raman analyses and "smart" imaging of nano-phases and nano-sized materials. *Spectrosc Eur*, 16:8–16, 2004.

[11] D. A. Long. *The Raman Effect- A Unified Treatment of the Theory of Raman Scattering by Molecules*. John Wiley & Sons Ltd, West Sussex PO19 1UD, England, 2002.

[12] M. Cardona and G. Gtintherodt. *Light Scattering In Solids VII*. Crystal-Field and Magnetic Excitations. Springer-Verlag, Berlin Heidelberg, 2000.

[13] W. Kaiser and M. Maier. *Stimulated Rayleigh, Brillouin, and Raman spectroscopy*, volume 2 of *Laser Handbook*. F.T.Arecchi and E.O.Schulz-Dubois, eds., North-Holland, Amsterdam, 1972.

[14] W.H. Weber and R. Merlin. *Raman Scattering in Material Science*. Springer Verlag, Berlin Heidelberg, 2000.

[15] W. J. Miller. *Symmetry groups and their applications*. ACADEMIC PRESS, New York, 1972.

[16] K. Dombrowski. *Micro-Raman Investigation of Mechanical Stress in Si Device Structures and Phonons in SiGe*. Dissertation, TU-Cottbus, 2000.

[17] W. Hayes. *Scattering of Light by Crystals*. John Wiley & Sons Inc, New York, 1 st edition, 1978.

[18] D. L. Rousseau, R. P. Bauman, and S. P. S. Porto. Normal Mode Determination in Crystals. *Journal of Raman Spectroscopy*, 10(Jan):253–290, 1981. doi:10.1002/jrs.1250100152.

[19] V. Senez, A. Armigliato, I. De Wolf, G. Carnevale, R. Balboni, S. Frabboni, and A. Benedetti. Strain determination in silicon microstructures by combined convergent beam electron diffraction, process simulation, and micro-Raman spectroscopy. *J. Appl. Phys.*, 94(9):5574–5583, 2003. doi:10.1063/1.1611287.

[20] S. M. Hu. Stress-related problems in silicon technology. *Journal of Applied Physics*, 70(6):R53–R80, 1991. doi:10.1063/1.349282.

[21] Park H. Jones K.S. Slinkman J.A. Law M.E. The effects of strain on dopant diffusion in silicon. *Proc. 1993. Int. Electron Devices Meeting (IEDM)*, pages 303–6, 1993. doi:10.1109/IEDM.1993.347347.

[22] I. Neizvestnyi and V. Gridchin. The use of stressed silicon in MOS transistors and CMOS structures. *Russian Microelectronics*, 38(2):71–86, 2009. doi:10.1134/S1063739709020012.

[23] D. W. Feldman, J. H. Parker, W. J. Choyke, and L. Patrick. Phonon Dispersion Curves by Raman Scattering in SiC Polytypes 3C,4H,6H,15R,and 21R. *Phys. Rev.*, 173(3):787–793, 1968. doi:10.1103/PhysRev.173.787.

[24] D. W. Feldman, J. H. Parker, W. J. Choyke, and L. Patrick. Raman Scattering in 6H SiC. *Phys. Rev.*, 170(3):698–704, 1968. doi:10.1103/PhysRev.170.698.

[25] J. Menendez and M. Cardona. Temperature-Dependence of the 1st-Order Raman-Scattering by Phonons in Si, Ge, and a-Sn - Anharmonic Effects. *Physical Review B*, 29(4):2051–2059, 1984. doi:10.1103/PhysRevB.29.2051.

[26] G. Lucazeau. Effect of pressure and temperature on Raman spectra of solids: anharmonicity. *J. Raman Spectrosc.*, 34(7-8):478–496, 2003. doi:10.1002/jrs.1027.

[27] S. Ganesan, A. A. Maradudin, and J. Oitmaa. A lattice theory of morphic effects in crystals of the diamond structure. *Annals of Physics*, 56(2):556–594, 1970. doi:10.1016/0003-4916(70)90029-1.

[28] E. Anastassakis, A. Pinczuk, E. Burstein, F. H. Pollak, and M. Cardona. Effect of static uniaxial stress on the Raman spectrum of silicon. *Solid State Communications*, 8(2):133–138, 1970. doi:10.1016/0038-1098(70)90588-0.

[29] Xiaoming Wu, Jianyuan Yu, Tianling Ren, and Litian Liu. Micro-Raman spectroscopy measurement of stress in silicon. *Microelectronics Journal*, 38(1):87–90, 2007. doi:10.1016/j.mejo.2006.09.007.

[30] A. B. Horsfall, J. M. M. d. Santos, S. M. Soare, N. G. Wright, A. G. O'Neill, S. J. Bull, A. J. Walton, A. M. Gundlach, and J. T. M. Stevenson. Direct measurement of residual stress in sub-micron interconnects. *Semicond. Sci. Technol.*, 18(11):992–996, 2003. doi:10.1088/0268-1242/18/11/315.

[31] S. Kouteva-Arguirova, T. Arguirov, D. Wolfframm, and J. Reif. Influence of local heating on micro-Raman spectroscopy of silicon. *Journal of Applied Physics*, 94(8):4946–4949, 2003. doi:10.1063/1.1611282.

[32] B. A. Weinstein and G. J. Piermarini. Raman-Scattering and Phonon Dispersion in Si and Gap at very High-Pressure. *Physical Review B*, 12(4):1172–1186, 1975. doi:10.1103/PhysRevB.12.1172.

[33] T. Beechem, S. Graham, S. P. Kearney, L. M. Phinney, and J. R. Serrano. Simultaneous mapping of temperature and stress in microdevices using micro-Raman spectroscopy. *Review of Scientific Instruments*, 78(6), 2007. doi:10.1063/1.2738946.

[34] C. Ulrich, A. Debernardi, E. Anastassakis, K. Syassen, and M. Cardona. Raman linewidths of phonons in Si, Ge, and SiC under pressure. *Physica Status Solidi B-Basic Research*, 211(1):293–300, 1999. doi:10.1002/(SICI)1521-3951(199901)211:1<293::AID-PSSB293>3.0.CO;2-O.

[35] J. Liu and Y. K. Vohra. Raman Modes of 6H Polytype of Silicon-Carbide to Ultrahigh Pressures - a Comparison with Silicon and Diamond. *Physical Review Letters*, 72(26):4105–4108, 1994. `doi:10.1103/PhysRevLett.72.4105`.

[36] J. Liu and Y. K. Vohra. Raman modes of 6H polytype of silicon carbide to ultrahigh pressures - Reply. *Physical Review Letters*, 77(8):1661–1661, 1996.

[37] A. Debernardi and M. Cardona. Dependence of phonon linewidths in semiconductors on temperature and isotopic composition. *Nuovo Cimento D Serie*, 20(7-8):923–930, 1998.

[38] A. Debernardi, C. Ulrich, K. Syassen, and M. Cardona. Raman linewidths of optical phonons in 3C-SiC under pressure: First-principles calculations and experimental results. *Physical Review B*, 59(10):6774–6783, 1999. `doi:10.1103/PhysRevB.59.6774`.

[39] A. Debernardi, C. Ulrich, M. Cardona, and K. Syassen. Pressure dependence of Raman linewidth in semiconductors. *Physica Status Solidi B-Basic Research*, 223(1):213–223, 2001. `doi:10.1002/1521-3951(200101)223:1<213::AID-PSSB213>3.0.CO;2-I`.

[40] D. Olego and M. Cardona. Temperature-Dependence of the Optical Phonons and Transverse Effective Charge in 3C-SiC. *Physical Review B*, 25(6):3889–3896, 1982. `doi:10.1103/PhysRevB.25.3889`.

[41] R. Jalilian, G. U. Sumanasekera, H. Chandrasekharan, and M. K. Sunkara. Phonon confinement and laser heating effects in Germanium nanowires. *Physical Review B (Condensed Matter and Materials Physics)*, 74(15):155421, 2006. `doi:10.1103/PhysRevB.74.155421`.

[42] Z. Li and R. C. Bradt. Thermal-Expansion and Thermal-Expansion Anisotropy of SiC Polytypes. *Journal of the American Ceramic Society*, 70(7):445–448, 1987. `doi:10.1111/j.1151-2916.1987.tb05673.x`.

[43] R. Zallen and M. L. Slade. Influence of pressure and temperature on phonons in molecular chalcogenides: Crystalline As_4S_4 and S_4N_4. *Phys. Rev. B*, 18(10):5775–5798, Nov 1978. `doi:10.1103/PhysRevB.18.5775`.

[44] A. Sen, S.L. Chaplot, and R. Mittal. Effects of pressure and temperature on the vibronic as well as the thermodynamic properties of $LiYF_4$ and $LiYbF_4$. *Journal of Physics Condensed Matter*, 14:975–986, February 2002. `doi:10.1088/0953-8984/14/5/303`.

[45] R. Pässler. Moderate phonon dispersion shown by the temperature dependence of fundamental band gaps of various elemental and binary semiconductors

including wide-band gap materials. *Journal of Applied Physics*, 88(5):2570–2577, 2000. doi:10.1063/1.1287601.

[46] A. Compaan and H. J. Trodahl. Resonance Raman-Scattering in Si at Elevated-Temperatures. *Physical Review B*, 29(2):793–801, 1984. doi:10.1103/PhysRevB.29.793.

[47] M. Balkanski, R. F. Wallis, and E. Haro. Anharmonic Effects in Light-Scattering due to Optical Phonons in Silicon. *Physical Review B*, 28(4):1928–1934, 1983. doi:10.1103/PhysRevB.28.1928.

[48] S. A. Hambir, J. Franken, D. E. Hare, E. L. Chronister, B. J. Baer, and D. D. Dlott. Ultrahigh time-resolution vibrational spectroscopy of shocked molecular solids. *J. Appl. Phys.*, 81(5):2157–2166, 1997. doi:10.1063/1.364269.

[49] H. Harima, S. Nakashima, and T. Uemura. Raman-Scattering from Anisotropic LO-Phonon-Plasmon-Coupled Mode in N-Type 4H-SiC and 6H-SiC. *J. Appl. Phys.*, 78(3):1996–2005, 1995. doi:10.1063/1.360174.

[50] M. R. Abel, S. Graham, J. R. Serrano, S. P. Kearney, and L. M. Phinney. Raman thermometry of polysilicon microelectromechanical systems in the presence of an evolving stress. *Journal of Heat Transfer-Transactions of the Asme*, 129(3):329–334, 2007. doi:10.1115/1.2409996.

[51] R. Tsu and J. G. Hernandez. Temperature-Dependence of Silicon Raman Lines. *Applied Physics Letters*, 41(11):1016–1018, 1982. doi:10.1063/1.93394.

[52] J. W. Pomeroy, M. Kuball, H. Lu, W. J. Schaff, X. Wang, and A. Yoshikawa. Phonon lifetimes and phonon decay in InN. *Applied Physics Letters*, 86(22):223501, 2005. doi:10.1063/1.1940124.

[53] Leah Bergman, Michael D. Bremser, William G. Perry, Robert F. Davis, Mitra Dutta, and Robert J. Nemanich. Raman analysis of the configurational disorder in $Al_xGa_{1-x}N$ films. *Applied Physics Letters*, 71(15):2157–2159, 1997. doi:10.1063/1.119367.

[54] Olaf Hollricher Wolfram Ibach. *High Resolution Optical Microscopy*. WITec, Jungingen, 2002.

[55] E. Hering and R. Martin. *Photonik: Grundlagen, Technologie und Anwendung*. Springer, Berlin, Berlin, 2006.

[56] H. W. Lo and A. Compaan. Raman Measurements of Temperature during cw Laser-Heating of Silicon. *J. Appl. Phys.*, 51(3):1565–1568, 1980. doi:10.1063/1.327809.

[57] H. W. Lo and A. Compaan. Raman Measurement of Lattice Temperature during Pulsed Laser Heating of Silicon. *Physical Review Letters*, 44(24):1604, 1980. doi:10.1103/PhysRevLett.44.1604.

[58] P. E. Van Camp, V. E. Van Doren, and J. T. Devreese. Microscopic Screening and Phonon Dispersion of Silicon: Moment Expansion for the Polarizability. *Phys. Rev. Lett.*, 42(18):1224, 1979. doi:10.1103/PhysRevLett.42.1224.

[59] Wikipedia. Silicium - Wikipedia, Die freie Enzyklopädie, 2009. [accessed 8-December-2009]. Available from: http://de.wikipedia.org/w/index.php?title=Silicium&oldid=67689997.

[60] D. E. Aspnes and A. A. Studna. Dielectric functions and optical parameters of Si, Ge, GaP, GaAs, GaSb, InP, InAs, and InSb from 1.5 to 6.0 eV. *Phys. Rev. B*, 27(2):985, Jan 1983. doi:10.1103/PhysRevB.27.985.

[61] Kohji Mizoguchi and Shin ichi Nakashima. Determination of crystallographic orientations in silicon films by Raman-microprobe polarization measurements. *Journal of Applied Physics*, 65(7):2583–2590, 1989. doi:10.1063/1.342787.

[62] M. Bauer, A. M. Gigler, C. Richter, and R. W. Stark. Visualizing stress in silicon micro cantilevers using scanning confocal Raman spectroscopy. *Microelectron. Eng.*, 85(5-6):1443–1446, 2008. doi:10.1016/j.mee.2008.01.089.

[63] J. B. Casady and R. W. Johnson. Status of silicon carbide (SiC) as a wide-bandgap semiconductor for high-temperature applications: A review. *Solid-State Electr.*, 39(10):1409–1422, 1996. doi:10.1016/0038-1101(96)00045-7.

[64] J. Baliga. *Silicon Carbide Power Devices*. Worlds Scientific Publishing Co. Pte. Ltd., Singapore, 2005.

[65] J. S. Sullivan and J. R. Stanley. 6H-SiC photoconductive switches triggered at below bandgap wavelengths. *Ieee Transactions on Dielectrics and Electrical Insulation*, 14(4):980–985, 2007. doi:10.1109/TDEI.2007.4286537.

[66] Z. Li and R. C. Bradt. Thermal-Expansion of the Hexagonal (4H) Polytype of SiC. *J. Appl. Phys.*, 60(2):612–614, 1986. doi:10.1063/1.337456.

[67] O. Kordina. *Growth and characterisation of silicon carbide power device material*. Dissertation, Linköping University, 1994.

[68] Olle Kordina. General Properties of Silicon Carbide, 2009. [accessed 8-December-2009]. Available from: http://www.ifm.liu.se/matephys/AAnew/research/sicpart/kordina2.htm.

[69] J. C. Burton, L. Sun, M. Pophristic, S. J. Lukacs, F. H. Long, Z. C. Feng, and I. T. Ferguson. Spatial characterization of doped SiC wafers by Raman spectroscopy. *Journal of Applied Physics*, 84(11):6268–6273, 1998. doi:10.1063/1.368947.

[70] K. Karch, P. Pavone, W. Windl, O. Schütt, and D. Strauch. Ab initio calculation of structural and lattice-dynamical properties of silicon carbide. *Phys. Rev. B*, 50(23):17054 – 17063, Dec 1994. doi:10.1103/PhysRevB.50.17054.

[71] Z. Li and R. C. Bradt. Thermal-Expansion of the Cubic (3C) Polytype of SiC. *J. Mat. Science*, 21(12):4366–4368, 1986. doi:10.1007/BF01106557.

[72] H. Harima, T. Hosoda, and S. Nakashima. Temperature measurement in a silicon carbide light emitting diode by Raman scattering. *Journal of Electronic Materials*, 28(3):141–143, 1999. doi:10.1007/s11664-999-0003-4.

[73] Z. Li and R. C. Bradt. Thermal-Expansion of the Hexagonal (6H) Polytype of Silicon-Carbide. *Journal of the American Ceramic Society*, 69(12):863–866, 1986. doi:10.1111/j.1151-2916.1986.tb07385.x.

[74] J. C. Burton, L. Sun, F. H. Long, Z. C. Feng, and I. T. Ferguson. First- and second-order Raman scattering from semi-insulating 4H-SiC. *Physical Review B*, 59(11):7282–7284, 1999. doi:10.1103/PhysRevB.59.7282.

[75] S. Nakashima, T. Mitani, J. Senzaki, H. Okumura, and T. Yamamoto. Deep ultraviolet Raman scattering characterization of ion-implanted SiC crystals. *Journal of Applied Physics*, 97(12), 2005. doi:10.1063/1.1931039.

[76] M. Bauer, A.M. Gigler, A.J. Huber, R. Hillenbrand, and R.W. Stark. Temperature-depending raman line-shift of silicon carbide. *Journal of Raman Spectroscopy*, 40(12):1867–1874, 2009. doi:10.1002/jrs.2334.

[77] Alexander M. Gigler, Andreas J. Huber, Michael Bauer, Alexander Ziegler, Rainer Hillenbrand, and Robert W. Stark. Nanoscale residual stress-field mapping around nanoindents in SiC by IR s-SNOM and confocal Raman microscopy. *Optics Express*, 17(25):22351–22357, 2009. doi:10.1364/OE.17.022351.

[78] A. L. Espana, A. G. Joly, W. P. Hess, and J. T. Dickinson. Laser-Induced Damage of Calcium Fluoride. *Journal Name: Journal of Undergraduate Research; Journal Volume: 4*, page Medium: X, 2004.

[79] J. H. Apfel, E. A. Enemark, D. Milam, W. L. Smith, and M. J. Weber. Effects of barrier layers and surface smoothness on 150-ps, 1.064 μm laser damage of AR coatings on glass. page p. 255, United States, 1977. NIST Spec. Pub. 5009.

[80] V. Liberman, M. Rothschild, J. H. C. Sedlacek, R. S. Uttaro, A. Grenville, A. K. Bates, and C. Van Peski. Excimer-laser-induced degradation of fused silica and calcium fluoride for 193-nm lithographic applications. *Optics Letters*, 24(1):58–60, 1999. `doi:10.1364/OL.24.000058`.

[81] T. Feigl, J. Heber, A. Gatto, and N. Kaiser. Optics developments in the VUV - soft X-ray spectral region. *Nuclear Instruments & Methods in Physics Research Section a-Accelerators Spectrometers Detectors and Associated Equipment*, 483(1-2):351–356, 2002. `doi:10.1016/S0168-9002(01)00060-2`.

[82] W. Arens, D. Ristau, J. Ullmann, C. Zaczek, R. Thielsch, N. Kaiser, A. Duparre, O. Apel, R. Mann, K., H. Lauth, H. Bernitzki, J. Ebert, S. Schippel, and H. Heyer. Properties of fluoride DUV excimer laser optics: influence of the number of dielectric layers. In *Laser-Induced Damage in Optical Materials: 1999*, volume 3902, pages 250–259. SPIE, 2000. `doi:10.1117/12.379314`.

[83] M. Marsi, A. Locatelli, M. Trovo, R. P. Walker, M. E. Couprie, D. Garzella, L. Nahon, D. Nutarelli, E. Renault, A. Gatto, N. Kaiser, L. Giannessi, S. Gunster, D. Ristau, M. W. Poole, and A. Taleb-Ibrahimi. UV/VUV free electron laser oscillators and applications in materials science. *Surface Review and Letters*, 9(1):599–607, 2002. `doi:10.1142/S0218625X02001768`.

[84] D. Ristau, W. Arens, S. Bosch, A. Duparre, E. Masetti, D. Jacob, G. Kiriakidis, F. Peiro, E. Quesnel, and A. V. Tikhonravov. UV-optical and microstructural properties of MgF_2-coatings deposited by IBS and PVD processes. In *Advances in Optical Interference Coatings*, volume 3738, pages 436–445. SPIE, 1999.

[85] Korth Kristalle GmbH. CALCIUM FLUORIDE, 2009. [accessed 8-December-2009]. Available from: `http://korth.de/eng/503728952d091450d/503728952d09c361d.htm`.

[86] V. Liberman, M. Switkes, M. Rothschild, S. T. Palmacci, J. H. C. Sedlacek, D. E. Hardy, and A. Grenville. Long-term 193-nm laser irradiation of thin-film-coated CaF_2 in the presence of H_2O. In Bruce W. Smith, editor, *Optical Microlithography XVIII*, volume 5754, pages 646–654. SPIE, 2004. `doi:10.1117/12.601468`.

[87] V. Liberman, M. Rothschild, S. T. Palmacci, Ni. N. Efremow, J. H. C. Sedlacek, and A. Grenville. Ambient effects on the laser durability of 157-nm optical coatings. In Anthony Yen, editor, *Optical Microlithography XVI*, volume 5040, pages 487–498. SPIE, 2003. `doi:10.1117/12.497525`.

[88] L.Parthier. Immersion Lithography: a new challenge for CaF_2 quality. In *Optical Microlithography XVIII*, volume 5754, pages presented, but unpublished in Proc. SPIE, 2005.

[89] J. W. Schopf, A. B. Kudryavtsev, D. G. Agresti, T. J. Wdowiak, and A. D. Czaja. Laser-Raman imagery of Earth's earliest fossils. *Nature*, 416(6876):73–76, 2002. doi:10.1038/416073a.

[90] O. Beyssac, J.-N. Rouzaud, B. Goffé, F. Brunet, and C. Chopin. Graphitization in a high-pressure, low-temperature metamorphic gradient: a Raman microspectroscopy and HRTEM study. *Contributions to Mineralogy and Petrology*, 143(1):19–31, 2002. doi:10.1007/s00410-001-0324-7.

[91] O. Beyssac, B. Goffé, C. Chopin, and J. N. Rouzaud. Raman spectra of carbonaceous material in metasediments: a new geothermometer. *Journal of Metamorphic Geology*, 20(9):859–871, 2002. doi:10.1046/j.1525-1314.2002.00408.x.

[92] O. Beyssac, B. Goffé, J.-P. Petitet, E. Froigneux, M. Moreau, and J.-N. Rouzaud. On the characterization of disordered and heterogeneous carbonaceous materials by Raman spectroscopy. *Spectrochimica Acta Part A: Molecular and Biomolecular Spectroscopy*, 59(10):2267–2276, 2003. doi:10.1016/S1386-1425(03)00070-2.

[93] A. Cuesta, P. Dhamelincourt, J. Laureyns, A. Martinez-Alonso, and J. M. D. Tascon. Comparative performance of X-ray diffraction and Raman microprobe techniques for the study of carbon materials. *Journal of Materials Chemistry*, 8(12):2875–2879, 1998. doi:10.1039/a805841e.

[94] C. Bény-Bassez and J.N. Rouzaud. *Scanning Electron Microscopy*. AMF O'Hare (Ed.), SEM Inc., Chicago, 1985.

[95] M. Bauer, M. Bischoff, S. Jukresch, T. Hülsenbusch, A. Matern, A. Görtler, R. W. Stark, A. Chuvilin, and U. Kaiser. Exterior surface damage of calcium fluoride outcoupling mirrors for DUV lasers. *Opt. Express*, 17(10):8253–8263, 2009. doi:10.1364/OE.17.008253.

[96] Michael Bauer, Martin Bischoff, Thomas Hülsenbusch, Ansgar Matern, Robert W. Stark, and Norbert Kaiser. Onset of the optical damage in CaF_2 optics caused by deep-UV lasers. *Opt. Lett.*, 34(24):3815–3817, 2009. doi:10.1364/OL.34.003815.

[97] N. Beermann, H. Blaschke, H. Ehlers, D. Ristau, D. Wulff-Molder, S. Jukresch, A. Matern, C. F. Strowitzki, A. Görtler, M. B. Gäbler, and N. Kaiser. Long Term Tests of Resonator Optics in ArF Excimer Lasers. In *XVII International Symposium on Gas Flow and Chemical Lasers & High Power Lasers 2008*, Lisbon, Portugal, 2008. doi:10.1117/12.816706.

[98] D. H. Lindsley. Experimental Studies of Oxide Minerals. *Reviews in Mineralogy*, 25:69–106, 1991.

[99] L. V. Gasparov, D. B. Tanner, D. B. Romero, H. Berger, G. Margaritondo, and L. Forró. Infrared and Raman studies of the Verwey transition in magnetite. *Phys. Rev. B*, 62(12):7939–7944, Sep 2000. doi:10.1103/PhysRevB.62.7939.

[100] A. Wang, K. E. Kuebler, B. L. Jolliff, and L. A. Haskin. Raman spectroscopy of Fe-Ti-Cr-oxides, case study: Martian meteorite EETA79001. *American Mineralogist*, 89(5-6):665–680, 2004.

[101] H. C. Soffel and E. Appel. Domain structure of small synthetic titanomagnetite particles and experiments with IRM and TRM. *Physics of The Earth and Planetary Interiors*, 30(4):348–355, 1982. doi:10.1016/0031-9201(82)90042-5.

[102] M. Lewis. Some Experiments on Synthetic Titanomagnetites. *Geophysical Journal of the Royal Astronomical Society*, 16(3):295–310, 1968. doi:10.1111/j.1365-246X.1968.tb00224.x.

[103] T. Dörfer, W. Schumacher, N. Tarcea, M. Schmitt, and J. Popp. Quantitative mineral analysis using raman spectroscopy and chemometric techniques. *Journal of Raman Spectroscopy*, 9999(9999):n/a, 2009. doi:10.1002/jrs.2503.

[104] S. A. Gilder and M. Le Goff. Systematic pressure enhancement of titanomagnetite magnetization. *Geophys. Res. Lett.*, 35, 2008. doi:10.1029/2008GL033325.

[105] M. Marshall and A. Cox. Effect of Oxidation on the Natural Remanent Magnetization of Titanomagnetite in Suboceanic Basalt. *Nature*, 230(5288):28–31, 1971. doi:10.1038/230028a0.

[106] Gary D. Storrick. *Thermoremanent behaviour of small multidomain synthetic Magnetites*. Dissertation, University of Pittsburgh, 1993.

[107] B. J. Wanamaker and B. M. Moskowitz. Effect of Nonstoichiometry on the Magnetic and Electrical Properties of Synthetic Single Crystal $Fe_{2.4}Ti_{0.6}O_4$. *Geophys. Res. Lett.*, 21:983–986, 1994. doi:10.1029/94GL00877.

[108] Olga N. Shebanova and Peter Lazor. Raman spectroscopic study of magnetite ($FeFe_2O_4$): a new assignment for the vibrational spectrum. *J. Solid State Chem.*, 174(2):424–430, 2003. doi:10.1016/S0022-4596(03)00294-9.

7 Publications

7.1 Visualizing stress in silicon microcantilevers using scanning confocal Raman spectroscopy

M. Bauer, A.M. Gigler, C. Richter, R.W. Stark

Microelectronic Engineering, Volume 85, Issues 5 - 6, Pages 1443 - 1446
Proceedings of the Micro- and Nano-Engineering 2007 Conference - MNE 2007

Permanent weblink:
http://dx.doi.org/10.1016/j.mee.2008.01.089

Visualizing stress in silicon micro cantilevers using scanning confocal Raman spectroscopy

M. Bauer [a], A.M. Gigler [a], C. Richter [b], R.W. Stark [a],*

[a] *Center for NanoScience and Department of Earth and Environmental Sciences, Ludwig-Maximilians-Universität München, 80333 Munich, Germany*
[b] *NanoWorld Services GmbH, 91058 Erlangen, Germany*

Received 1 October 2007; received in revised form 14 January 2008; accepted 17 January 2008
Available online 8 February 2008

Abstract

We report on the determination of surface stress in a resonantly oscillating silicon micro cantilever using confocal Raman spectroscopy. Focusing on the optical phonon line of silicon allows one to determine the lateral distribution of mechanical stress. However, the Raman shift caused by mechanical stress can be concealed by thermally induced Raman shifts and line-broadenings. Both effects are intrinsic to the micro Raman measurement which uses a strongly focused laser beam in a confocal microscope. In order to unravel the different contributions, we use a practical analytical method for the compensation based on reference measurements on a heated silicon wafer of the same crystal orientation. As an example, the structure of the micro cantilever was specially weakened by introducing a rectangular hole in the center of the lever. After compensation of the thermally induced shift, the true mechanical stress can be visualized as shown for a stress maximum of the cantilever driven at its second flexural eigenmode.
© 2008 Elsevier B.V. All rights reserved.

Keywords: Silicon; Stokes shift; Stress; Raman spectroscopy; Temperature dependence; Local heating; Micro structure

1. Introduction

Confocal Raman microscopy allows to visualize the lateral distribution of stress in silicon microstructures [1,2]. Stress can be caused by fabrication (residual stress), temperature gradients, or mechanical loading. However, the measurement intrinsically heats the sample within the focus of the laser beam due to absorption of laser light [3,4]. This heating may conceal the actual stress induced line-shifts due to concurrent thermally induced line-shifts and line-broadenings. The peak position and width shows a linear dependence due to increasing temperature [5]. Here, we present a practical method to distinguish between the shift of the first order silicon Raman peak caused by mechanical stress or strain (both dynamic and static) and the influence of local sample heating. The benefit of this compensation strategy becomes obvious when dynamic bending stresses have to be investigated as in the case of resonantly driven microcantilevers. For homogeneous substrates without structure, the line shift due to local heating is constant. Due to heat conduction in the material, the heating shift causes a constant offset. In the case of structured samples, the heat conductance may vary locally, resulting in temperature gradients which cause thermal stress.

In Raman spectroscopy of solids, the inelastic scattering of a photon with molecular vibrations and crystal phonons is measured. Energy transfer can occur due to inelastic scattering: from the photon to a phonon (Stoke shift) or vice versa (anti-Stokes shift). Confocal Raman spectroscopy combines the advantages of confocal microscopy (high depth and diffraction limited spatial resolution) and Raman spectroscopy (chemical sensitivity). In case of crystal surfaces, Raman scattering depends on the phonon band structure of the lattice [6]. For silicon, scattering occurs with the optically coupled phonons near the center of the Brillouin zone [4] causing a shift of 521 cm^{-1} or 15.5 THz with respect to the excitation [6]. The effect of

* Corresponding author. Tel.: +49 89 2180 4329.
E-mail address: stark@lrz.uni-muenchen.de (R.W. Stark).

temperature on Raman line-widths has been discussed in the literature [7–9]. An increased temperature causes a distortion of the crystal lattice and, thus, different occupations of the phonon bands, i.e. a decrement of the phonon decay [5,10]. Furthermore, the equilibrium positions of the atoms are changed, resulting in a lattice variation. This changes phonon frequencies and causes an additional line-shift [11–13].

The mechanical contribution to the measured line-shift correlates with the elastic deformation of the crystal lattice due to mechanical loading. This line shift is mainly independent from temperature [14]. With lattice deformation, the phonon frequency is changed, but it should not influence the phonon life time or distribution (no line-broadening) [3]. Furthermore, line-broadening by mechanical stress is very small and can be neglected since the phonon population density remains more or less static [4]. This is the case for moderate temperatures and stresses [11]. Both mechanical and thermal stress causes a line-shift in the Raman spectrum of silicon. Depending on the direction of the stress field (tensile or compressive), these additional shifts can compensate or intensify the stress induced shift.

2. Materials and methods

All measurements were done using a confocal Raman microscope alpha300 R (www.witec.de; WITec, Jungingen, Germany) with a piezo scan stage (100 × 100 × 20 μm, PI, Germany). The system was equipped with a 100× microscope objective for measuring in air with a working distance of 0.26 mm and a numerical aperture NA = 0.90 (Nikon, Düsseldorf, Germany). The depth of focus was about 1 μm. The microscope objective was used to focus the excitation beam of a 2ω–Nd:YAG laser (532 nm emission) onto the surface and to collect the scattered light. The zero-th order light was blocked by a long-pass edge filter for 532 nm. To achieve a confocal setup, the collected light was focused onto the core of a 50 μm multimode fiber acting as the pin hole and guiding the light to the lens based spectrometer. In the experiments discussed here, we used a 500 nm-blazed diffraction grating with 1800 lines/mm. A back thinned, peltier cooled CCD-Chip (1024 × 128 pixel, cooled to −65 °C) was used to record the spectra at a minimum integration time of 19 ms.

A custom made sample stage with a peltier element (Conrad TEC1-1703, 3.9 W) was used to control the sample temperature in the range of 10–150 °C. The temperature was measured directly underneath the sample by a PT 100 element connected to a temperature controller (ST70-31.03, Störk, Germany, 0.1 K accuracy).

For calibration purposes, a polished silicon (1 0 0) wafer was used. The dynamic bending measurements were obtained for a silicon cantilever as used in an atomic force microscope (AFM). The cantilever had a rectangular weakening in the central region of the lever. At each point of the image, a complete spectrum was recorded. Within a preset window from 470 cm^{-1} to 550 cm^{-1}, a Gaussian fit was conducted resulting in a map of peak width and center position of the peak.

Single spectra (0.5 s integration time) were obtained on a silicon wafer at a given sample position after thermal equilibration. After each spectrum, the temperature was increased by 5 K. These single spectra were then processed using a Gaussian curve fit (using Microcal Origin 7.5 pro). Both, Stokes peak center v and peak width W depend on the temperature T. The relations between line shift and temperature and peak width and temperature are linear at moderate temperatures [4]. Thus, the results for the peak center v and peak width W were plotted versus temperature T and linearly fitted with the equations

$$v(T) = m_{center} \cdot T + c_{center}, \quad (1)$$

$$W(T) = m_{therm} \cdot T + c_{therm}. \quad (2)$$

Here, the parameters m and c are the respective fitting parameters. The total line shift is caused by a combination of stress and heating. In order to calculate the line shift induced by mechanical stress Δv_{stress}, the thermal contribution has to be subtracted

$$\Delta v_{stress} = v - v_{therm}. \quad (3)$$

By inserting Eq. (1) and (2) into Eq. (3) we obtain an expression for Δv_{stress} which only contains the measured center of the Stokes line v, fitted peak width W and the fitting parameters m and c

$$\Delta v_{stress} = v - \left[\frac{m_{center}}{m_{width}}(W - c_{therm}) + c_{center}\right] \quad (4)$$

3. Results and discussion

The dependence of the center position and line-width of the Stokes peak of silicon on the laser power is shown in Fig. 1. A piece of silicon wafer without external stress was scanned, while the laser power was increased every 20 lines from 1 μW to a maximum of 22.3 mW. Simultaneously with the increase in laser power, the peak position is shifting to lower wave numbers and the line-width is

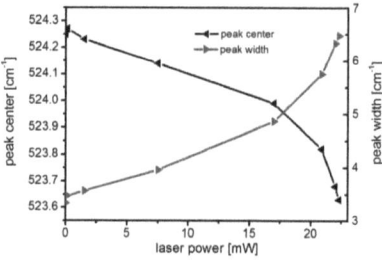

Fig. 1. Effect of increasing laser power on the Stokes line shape. The average line position and width are plotted versus the laser power (more than 3500 spectra per data point).

increasing (Fig. 1). Each point resembles the average about 3500 spectra. Peak position and peak width are clearly anti-correlated and caused by the increasing local heating with increasing laser power.

In a reference experiment, a piece of silicon wafer was heated from 20 °C to 100 °C in steps of 5 K. At each temperature step, a spectrum was recorded. Peak position and peak width are plotted versus sample temperature in Fig. 2. For both parameters, a strictly linear behavior is observed within the accessible temperature range. The slopes of the linear fits were used for the temperature compensation calculations. We obtain $m_{center} = -0.025$ K/cm^{-1} for the shift of the peak center and $m_{width} = 4.30$ K/cm^{-1} for the change in line-width. These results support values reported in the literature [3,4], even though we were using a stronger focus of the excitation laser. Thus, the compensation equation is

$$\Delta v_{stress} = v - \left[\frac{-0.025}{0.01}(W - 4.30) + 522.02\right] \quad (5)$$

Thus, the stress induced line shift can be calculated from the measured Raman lines peak position v and its peak width W on silicon devices.

Confocal Raman images were obtained on a specially modified AFM micro cantilever (Fig. 3a) which was driven at its second flexural eigenmode (463 kHz). The focus was adjusted slightly above the silicon surface. Thus, the vibrating structure was in-focus at the upper turning point and out-of focus at the lower turning point. During scanning over the mechanical structure, local temperature gradients can be formed. Possible reasons are defocusing of the laser spot due to a slight misalignment between structure and focal plane as well as anisotropic heat conductance within the thin structure, e.g. close to the hole. Therefore, mechanical stress cannot be visualized by the sole confocal Raman measurement. Fig. 3a shows the center position of the optical phonon line as obtained by a Gaussian fit to the measured spectra. The characteristic stress distribution of the second eigenmode is hidden by thermally induced line shift (Fig. 3b, right). As a first step in the compensation process,

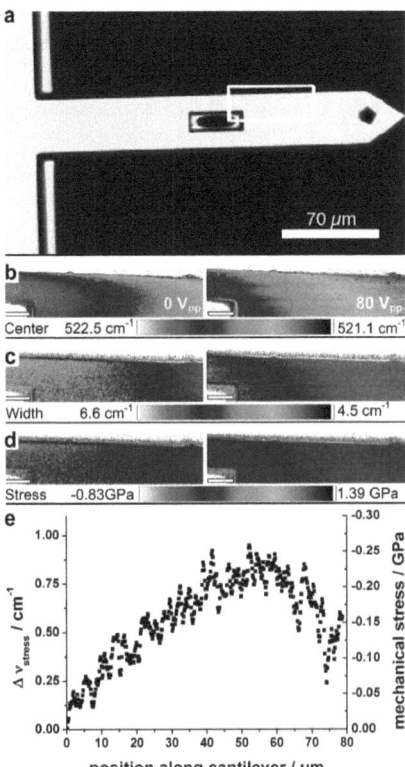

Fig. 3. Optical and confocal Raman images of the modified region of an AFM cantilever, resonantly driven at its second eigenmode (463 kHz) at 0 V$_{pp}$ and 80 V$_{pp}$ (scale-bar: 10 μm); (a) optical micrograph (20×) of the silicon micro cantilever. The rectangle indicates the region of interest, (b) Uncompensated Gaussian peak center, (c) corresponding peak width, (d) temperature compensated stress map and (e) cross section through the stress distribution.

the peak width was calculated as shown in Fig. 3c. Upon compensation of the thermally induced shift, the development of the stress was calculated (Fig. 3d). For the in-plane stress on a Si (100) surface along the [100] direction, the stress varies linearly with Stokes peak center with the proportionality constant $D = -3.6$ cm^{-1}/GPa [4]. Compared to the un-driven cantilever that may be subject to residual stress from the manufacturing process, the oscillation at the second flexural resonance causes an accumulation of stress at the maximum of oscillation upon stronger agitation of the cantilever. A cross sectional analysis in Fig. 3e reveals a stress minimum due to the node of the second flexural eigenmode.

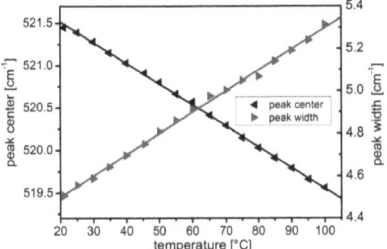

Fig. 2. Temperature dependence of Stokes peak center and width on a heated Si-wafer. Each data point was calculated from a Gaussian fit. The solid line shows the corresponding linear fit. These fitting results were used for the temperature compensation.

Finally, it should be mentioned that with our simple 'defocusing method', one obtains a reasonable picture of the stress distribution within the structure while the numerical values might underestimate the true stresses. In our experiment we average over the entire mechanical oscillation cycle which implies both tensile and compressive stress. By slightly defocusing, we have emphasized the part of the oscillation cycle where compressive stress dominates.

4. Conclusion

In confocal Raman microscopy of silicon microstructures, additional line-shifts are induced by both local heating due to the focused laser beam and mechanical stress. Both effects, line shift due to heating and due to stress can be on the same order of magnitude making a direct stress measurement impossible. Temperature effects can be removed by a simple fitting procedure which is based on the line broadening due to heating. Thus, it was possible to visualize the dynamical bending stresses in a silicon AFM micro cantilever which was driven at the second flexural eigenmode. The present results clearly indicate that the characterization of the stress distribution in a vibrating micro structure is accessible by micro Raman spectroscopy. In order to achieve a more quantitative data during the different stages of vibration, a more elaborate laser illumination is required. We suggest that by adequately chopping the laser, it will be possible to investigate all stages of vibration.

Acknowledgements

We gratefully acknowledge financial support by European Commission under contract NMP4-CT-2004-013684 (Force Tool).

References

[1] I. De Wolf, Semicond. Sci. Technol. 11 (2) (1996) 139–154.
[2] V.I. Srikar, A.K. Swan, M.S. Unlu, B.B. Goldberg, S.M. Spearing, J. Microelectromech. Syst. 12 (6) (2003) 779–787.
[3] S. Kouteva-Arguirova, T. Arguirov, D. Wolfframm, J. Reif, J. Appl. Phys. 94 (8) (2003) 4946–4949.
[4] T. Beechem, S. Graham, S.P. Kearney, L.M. Phinney, J.R. Serrano, Rev. Sci. Instrum. 78 (6) (2007) 061301.
[5] R. Tsu, J.G. Hernandez, Appl. Phys. Lett. 41 (11) (1982) 1016–1018.
[6] K. Dombrowski, edited by TU-Cottbus (2000).
[7] S.A. Hambir, J. Franken, D.E. Hare, E.L. Chronister, B.J. Baer, D.D. Dlott, J. Appl. Phys. 81 (5) (1997) 2157–2166.
[8] D.E. Hare, J. Franken, D.D. Dlott, J. Appl. Phys. 77 (11) (1995) 5950–5960.
[9] M.R. Abel, S. Graham, J.R. Serrano, S.P. Kearney, L.M. Phinney, J. Heat Transfer 129 (3) (2007) 329–334.
[10] A. Compaan, H.J. Trodahl, Phys. Rev. B 29 (2) (1984) 793–801.
[11] G. Lucazeau, J. Raman Spectrosc. 34 (7–8) (2003) 478–496.
[12] J. Menendez, M. Cardona, Phys. Rev. B 29 (4) (1984) 2051–2059.
[13] T.R. Hart, R.L. Aggarwal, B. Lax, Phys. Rev. B 1 (2) (1970) 638–642.
[14] B.A. Weinstein, G.J. Piermarini, Phys. Rev. B 12 (4) (1975) 1172–1186.

7.2 Temperature depending Raman line-shift of silicon carbide

Michael Bauer, Alexander M. Gigler, Andreas J. Huber, Rainer Hillenbrand and Robert W. Stark

Journal of Raman Spectroscopy, 2009, Volume 40 Issue 12, Pages 1867-1874

Permanent weblink:
http://dx.doi.org/10.1002/jrs.2334

Temperature-depending Raman line-shift of silicon carbide

Michael Bauer,[a,b] Alexander M. Gigler,[a,b] Andreas J. Huber,[b,c] Rainer Hillenbrand[b,c] and Robert W. Stark[a,b]*

Silicon carbide (SiC) is often used for electronic devices operating at elevated temperatures. Spectroscopic temperature measurements are of high interest for device monitoring because confocal Raman microscopy provides a very high spatial resolution. To this end, calibration data are needed that relate Raman line-shift and temperature. The shift of the phonon wavenumbers of single crystal SiC was investigated by Raman spectroscopy in the temperature range from 3 to 112 °C. Spectra were obtained in undoped 6H–SiC as well as in undoped and nitrogen-doped 4H–SiC. All spectra were acquired with the incident laser beam oriented parallel as well as perpendicular to the c-axis to account for the anisotropy of the phonon dispersion. Nearly all individual peak centers were shifting linearly towards smaller wavenumbers with increasing temperature. Only the peak of the longitudinal optical phonon A_1(LO) in nitrogen-doped 4H–SiC was shifting to larger wavenumbers. For all phonons, a linear dependence of the Raman peaks on both parameters, temperature and phonon frequency, was found in the given temperature range. The linearity of the temperature shift allows for precise spectroscopic temperature measurements. Temperature correction of Raman line-shifts also provides the ability to separate thermal shifts from mechanically induced ones. Copyright © 2009 John Wiley & Sons, Ltd.

Keywords: silicon carbide; temperature; Raman line-shift

Introduction

Silicon carbide (SiC) is a semiconductor that can be used for high-temperature and high-voltage applications. Local temperature and stress measurements are needed for the optimization of fabrication processes and to monitor the devices during operation. Raman spectroscopy is a powerful tool for such measurements in crystalline materials by analyzing line-shift and shape of the Raman peaks.[1,2] A shift of the phonon peaks can be induced by temperature changes or mechanical stress. Thus, a precise temperature calibration allows one to separate both effects and to achieve compensated values for local temperature and stress measurements under typical operational conditions of about −20 to +120 °C.

Raman spectra of crystalline materials depend on the inelastic scattering of light at phonons and consequently on the phonon band structure of the lattice.[3] This relationship can give direct access to the phonon band structure of SiC single crystals.[4,5] SiC polytypes are named according to the number n of double layers in their unit cell followed by the lattice type (C, H, R) with cubic 3C–SiC being the most basic structure. For each polytype, the length of the unit cell along the c-axis is n times larger than that of the basic 3C–SiC type. Due to this increased unit cell, the Brillouin zone is decreased in the Γ–L direction by 1/n coinciding with a reduced wavevector ($x = q/q_B$) of the phonon mode in the basic Brillouin zone ($q = 0$). For higher order polytypes, the dispersion curves of the propagating phonons are folded backwards into the basic zone creating additional phonon modes, usually referred to as folded modes.[6,7]

Raman spectroscopy can also be used to measure the distribution of dopants within polar semiconductors.[6] Depending on the nitrogen concentration in SiC, width, spectral shape, and peak position of the longitudinal optical phonon A_1(LO) phonon change. These variations have been attributed to plasmon–phonon coupling.[6,8] Nitrogen doping of SiC generates free carriers in the semiconductor. Light can excite oscillations of these free carriers (plasmons) that can couple to the longitudinal optical phonons leading to a plasmon–phonon coupled LO mode (LOPC). The frequency and line-shape of the LOPC peak can be described analytically based on the dielectric function.[6,9] The LOPC line allows one to measure carrier density and mobility.[10–13] Its line-shift can be used to characterize ion-implanted SiC with different annealing temperatures to deduce the free-carrier density[13,14] and the spatial distribution of the dopants.[6]

Increasing temperature causes a distortion of the lattice and thus, a change in population of the phonon bands that results in relative line-shifts and line-broadening.[15,16] Due to the linear expansion of the equilibrium position of the atoms, the bond strength changes for moderate temperatures.[17] This modifies the phonon wavenumbers and leads to a shift of the corresponding Raman peaks. The anisotropy of thermal expansion of SiC shows a non-linear behavior for high temperatures giving rise to varying lattice parameters as well as unit cell vectors.[18,19]

The temperature-induced shift of the peak position in Raman spectra can be used for local temperature measurements as it was

* Correspondence to: Robert W. Stark, Department of Earth and Environmental Sciences, Ludwig-Maximilians-Universität München, Theresienstr. 41, 80333 Munich, Germany. E-mail: stark@lrz.uni-muenchen.de

a Department of Earth and Environmental Sciences, Ludwig-Maximilians-Universität München, Theresienstr. 41, 80333 Munich, Germany

b Center for NanoScience (CeNS), Schellingstr. 4, 80799 Munich, Germany

c Nanooptics Laboratory, CIC nanoGUNE, 20009 Donostia – San Sebastian, Spain

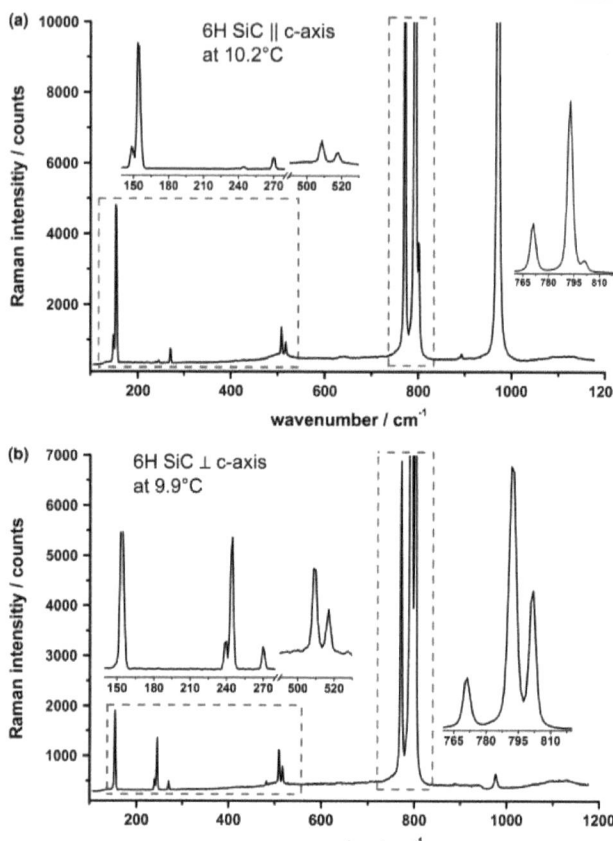

Figure 1. Raman spectra of 6H–SiC. (a) Spectrum measured with the incident beam parallel and (b) perpendicular to the c-axis. This figure is available in colour online at www.interscience.wiley.com/journal/jrs.

demonstrated for a light-emitting SiC diode using the shift of the E_2 Folded Transversal Optical (FTO) Raman peak of 6H–SiC.[2] For this peak, the temperature-dependent Raman peak center was approximated by the following:

$$\omega(T) = \omega_0 - C\left(1 + \frac{2}{e^x - 1}\right) - D\left(1 + \frac{3}{e^y - 1} + \frac{3}{(e^y - 1)^2}\right) \quad (1)$$

with

$$x = \frac{\hbar\omega_0}{2k_B T} \quad (2)$$

and

$$y = \frac{\hbar\omega_0}{3k_B T} \quad (3)$$

Here, $k_B T$ is the thermal energy. In $\hbar\omega_0$, the anharmonic interaction is neglected. C and D are free fitting parameters. For temperatures less than 500 °C, the third term in Eqn (1) is negligible which leads to a linear temperature dependence. At higher temperatures (up to 1000 °C), four anharmonic terms have to be considered in a Hamiltonian model to describe the phonon correctly.[2]

In addition to temperature, mechanical stress can influence the peak position due to lattice distortion. This effect was investigated in both experiment and theory.[20] In silicon devices, for example, a similar dependence of the peak center and peak width on stress and temperature has been observed. In silicon, the incident photons are scattered at the optical phonons near the center of the Brillouin zone at a frequency of 15.5 THz corresponding to the first-order Raman line at 521 cm^{-1}. The temperature dependence of this Raman peak is well documented in literature.[15,21,22] A linear relation with temperature has been reported for both peak position and line-width.[23] This linearity can be used for temperature and stress measurements in silicon microdevices.[1] However, due to the high absorption coefficient of 8.24×10^3 per cm^{-1} in silicon at 532 nm, the laser beam used for the temperature measurement

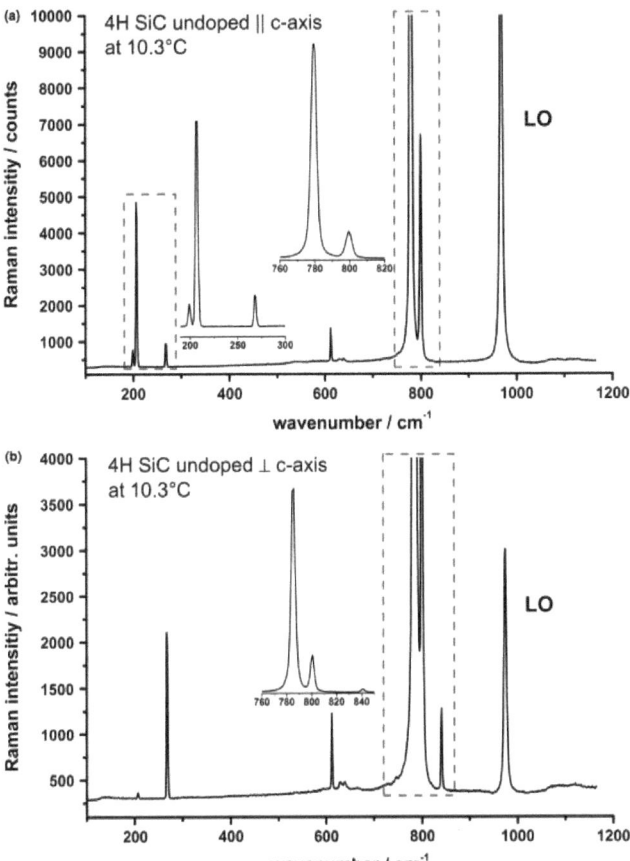

Figure 2. Raman spectra of undoped 4H–SiC. (a) Spectrum measured with the incident beam parallel to the c-axis and (b) perpendicular to the c-axis. This figure is available in colour online at www.interscience.wiley.com/journal/jrs.

causes changes in the Raman spectra by local heating, which have to be taken into consideration.[1,24]

The spectral position and line-width of the longitudinal optical phonons in SiC increase with pressure, whereas the transverse optical (TO) phonons remain unchanged.[25] Temperature and stress effects are relevant for the design and operation of SiC microdevices, which are usually fabricated out of several different materials, thus, introducing different thermal expansion coefficients into the compound. As a consequence, mechanical stress can occur within a semiconductor device on temperature change. Depending on the direction of mechanical stress (compressive or tensile), thermally and mechanically induced peak shifts may compensate or intensify each other.[26] However, both effects can be separated because they have disparate influences on the phonons. Moderate mechanical stress only influences the peak position, whereas temperature affects both peak position and line-width. Thus, temperature correction factors can be calculated from an experiment with controlled temperature. Based on this calibration, Raman data can be used for non-contact temperature measurements. In the following, we report on measurements of the temperature shift of Raman peaks for the polytypes 6H–SiC (undoped) and 4H–SiC (doped and undoped). We used the fit parameters of Eqn (1) to determine the thermal dependency of the Raman shifts.

Materials and Methods

Pieces of polished SiC wafers with a size of about $1 \times 1\,cm^2$ and 1 mm thickness were analyzed. The samples were cut from undoped 6H–SiC and 4H–SiC single crystals as well as from a nitrogen-doped 4H–SiC wafer (doping concentration: $5 \times 10^{18}\,cm^{-3}$).

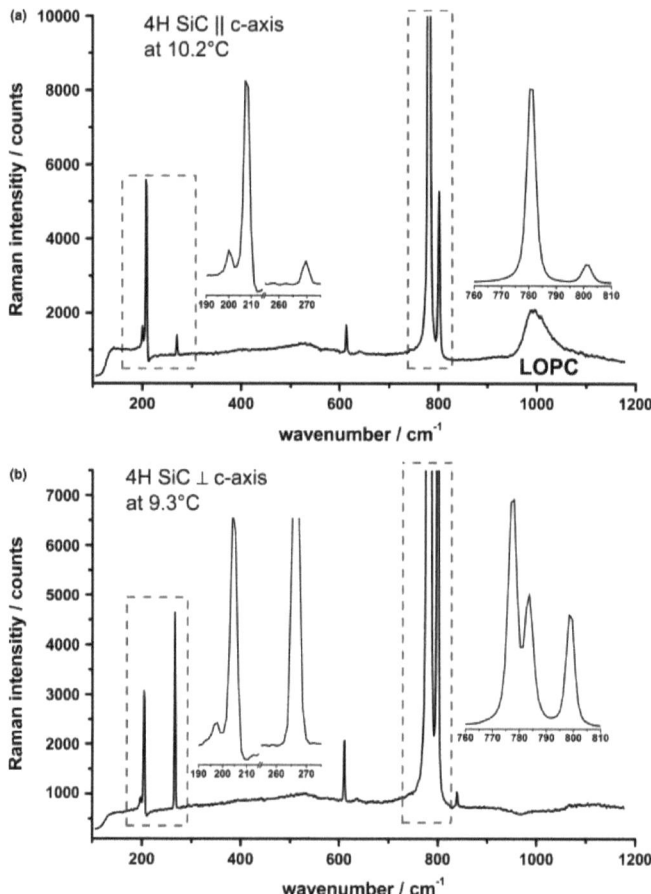

Figure 3. Raman spectra of nitrogen-doped 4H–SiC. (a) Spectrum measured with the incident beam parallel to the c-axis and (b) perpendicular to the c-axis. In both, details illustrate weaker Raman peaks. This figure is available in colour online at www.interscience.wiley.com/journal/jrs.

The spectra were measured with a confocal Raman microscope (alpha300 R; WITec GmbH, Ulm, Germany). This system employed the 50 µm core of a multimode fiber as pinhole in backscattering geometry. For excitation, a Nd:YAG SHG 532 nm laser with a maximum power output of 22.5 mW was used together with a 100× objective with a working distance of 0.26 mm and a numerical aperture of 0.90. With this setup, the focal depth was limited to 1 µm. Elastically scattered photons (Rayleigh scattering and reflexion) were rejected by a sharp edge filter. Spectra were acquired by a lens-based spectrometer with a CCD-camera (1024 × 128 pixel cooled to −65 °C). The spectral resolution of the setup was 1.17 cm^{-1} per CCD-pixel for a 1800 mm^{-1} grating. Using a Lorentzian to fit Raman peaks, a nominal sub-pixel resolution of 0.02 cm^{-1} could be achieved. The Raman scattering process for SiC was non-resonant due to the band gap of 3.3 eV for 4H–SiC and 3.0 eV for 6H–SiC with respect to the photon energy of 2.33 eV.

Temperature control was realized by Peltier elements for cooling and heating (Conrad TEC1-1703, 3.9 W). The temperature sensor (PTC element, accuracy: ±0.1 K) was located directly beneath the sample. For the experiments, the SiC samples were first cooled to the lowest achievable temperature (3 °C) and then heated in steps of 5 K (±1 K) up to 112 °C.

A single spectrum with an integration time of 7 s was recorded at each temperature step. The optical system was refocused for each measurement to compensate for thermal expansion or drift of the measurement system. To account for the anisotropy of Raman scattering, all samples were oriented with the laser parallel to the c- and a-axis (perpendicular to the c-axis) of the SiC crystal. Peak center position and full width at half maximum (FWHM) of

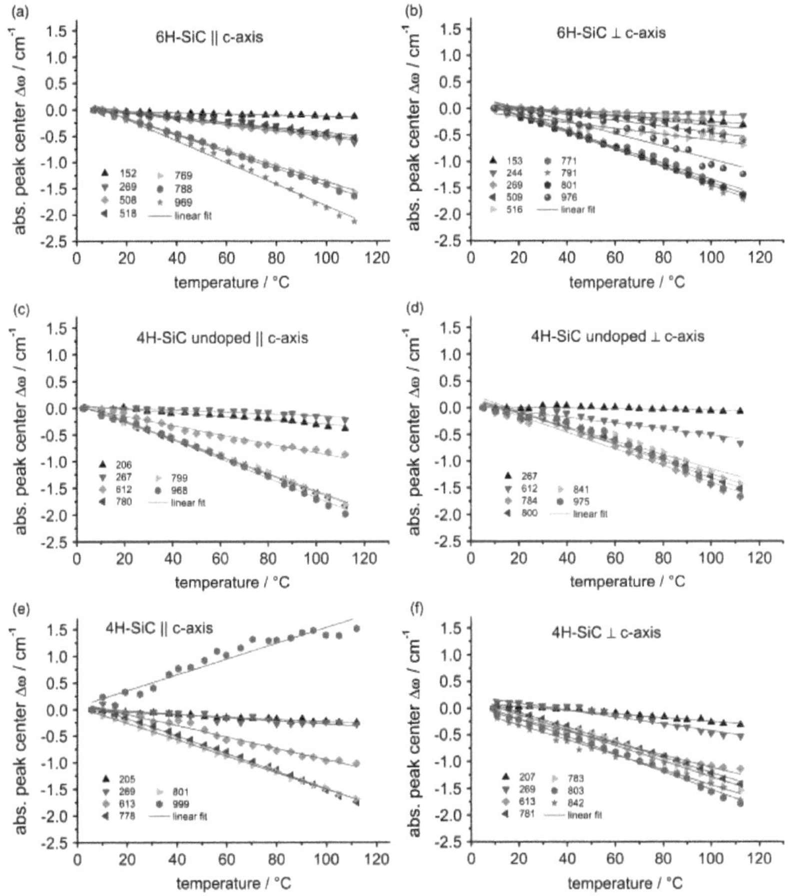

Temperature-depending Raman line-shift of silicon carbide

Figure 4. Absolute temperature shifts $\Delta\omega = (\omega_c - \omega_{c0})$ of the Raman lines. Raman peaks of 6H–SiC measured with the laser oriented (a) parallel and (b) perpendicular to the c-axis. (c and d) Raman peaks of undoped 4H–SiC in parallel and perpendicular orientation and (e and f) nitrogen-doped 4H–SiC in both orientations. This figure is available in colour online at www.interscience.wiley.com/journal/jrs.

prominent Raman lines were obtained from a Lorentzian curve fit using Microcal origin.

Results and Discussion

Raman spectra of the polytypes

The laser power was changed stepwise from 0.1 mW to a maximum power of 22.5 mW to quantify the local heating of the SiC samples by the incident beam. For each temperature step, a single spectrum was acquired by averaging over five accumulations, with 7 s integration time each. All Raman peaks were fitted by a Lorentzian curve. The change in peak center position due to laser heating was on the same order of magnitude for all polytypes and peaks

(5.3 cm^{-1}/W for 6H–SiC and 6.1 cm^{-1}/W for 4H–SiC). Because – for 6H–SiC and 4H–SiC – the absorption coefficient ($\alpha = 1.59$ per cm at 532 nm) is very small and thus, the penetration depth[11] ($\delta \sim 3$ mm) is much larger than the focal depth of 1 µm, only a very small fraction of the laser energy is absorbed in the material. For this reason, heating effects by the laser can be neglected in confocal Raman microscopy of SiC. For comparison, a similar measurement on silicon revealed a much larger peak shift of 30 cm^{-1}/W, which made compensation strategies necessary.[26]

Raman spectra for different crystal orientations are shown in Figs 1–3. The spectra were measured at 10 °C sample temperature. Insets show details of selected peaks. Depending on the orientation between the crystal and the laser beam, the Raman peaks showed splitting, changed in scattering intensity, or occurred

Table 1. Peak center positions of the Raman peaks measured in undoped 4H–SiC and 6H–SiC and nitrogen-doped 4H–SiC[a]

Polytype	$x = q/q_b$	FTA, transversal (planar) acoustic				FLA, longitudinal (axial) acoustic			
		$\parallel$ (cm^{-1})		$\perp$ (cm^{-1})		$\parallel$ (cm^{-1})		$\perp$ (cm^{-1})	
4H undoped	0	–		–		–		–	
	2/4	199.1, 207.2	E_2	206.6	E_2	–		–	
	4/4	268.1	E_2	268.1	E_1	612.5	A_1	612.2	A_1
4H doped	0	–		–		–		–	
	2/4	200.0, 208.1	E_2	196.8, 204.6	E_2	–		–	
	4/4	269.5	E_1	266.4	E_2	613.9	A_1	610.6	A_1
6H undoped	0	–		–		–		–	
	2/6	148.5, 154.2	E_2	153.9	E_2	–		481.8	B_1
	4/6	244.8	E_2	239.4, 244.3	E_1	508.5, 517.6	A_1	509.2, 516.8	A_1
	6/6	270.3	E_1	270.2	E_2	–		–	

Polytype	$x = q/q_b$	FTO, transversal (planar) optic				FLO, longitudinal (axial) optic			
		$\parallel$ (cm^{-1})		$\perp$ (cm^{-1})		$\parallel$ (cm^{-1})		$\perp$ (cm^{-1})	
4H undoped	0	799.4	E_2	800.3	E_1	968.3	A_1	974.6	A_1
	2/4	779.4	E_2	785.4	E_2	–		–	
	4/4	–		–		–		840.9	A_1
4H doped	0	801.2	E_1	798.9	E_1	999.2	A_1	–	
	2/4	780.9	E_2	777.5, 783.1	E_2	–		–	
	4/4	–		–		–		839.4	A_1
6H undoped	0	800.7	E_1	801.4	E_1	970.4	A_1	976.4	A_1
	2/6	792.7	E_2	792.2	E_2	–		–	
	4/6	–		–		893.2	A_1	–	
	6/6	771.0	E_2	770.9	E_2	–		–	

[a] The assignment of the Raman peaks with phonon modes and symmetries follows Ref. [7].

exclusively in one orientation. Due to the finite angle between the crystal axis and the direction of the incident photons, Raman scattering also occurred at phonons corresponding to other crystal orientations. This explains small additional peaks in the spectra. The assignment and origin of the Raman lines are given in Table 1.

Undoped 6H–SiC

The Raman spectrum of 6H–SiC with the laser beam oriented in parallel alignment with the c-axis is shown in Fig. 1(a). Figure 1(b) shows the Raman spectrum of the same sample, measured with the incident laser perpendicular to the c-axis. Both measurements were acquired at about 10 °C. The major peaks were nearly at the same spectral position for both orientations. The single peak at 154 cm^{-1} (Folded Transversal Acoustic (FTA) mode, E_2 symmetry) measured for parallel orientation, however, was split into two peaks in the spectrum recorded in perpendicular orientation. In parallel orientation, the peaks at 240 cm^{-1} (split FTA mode, E_1 symmetry) were not as intense as in the perpendicular orientation. For 509 cm^{-1} (Folded Longitudinal Acoustic (FLA), A_1 symmetry) and 517 cm^{-1} (FLA, A_1 symmetry) in the 6H–SiC spectra – present in both orientations – an origin from Si-Si nearest neighbors could be assumed.[6] This effect will be discussed later. Note that the Si-Si Raman band appears at 521 cm^{-1} for unstressed silicon crystals. For the folded longitudinal acoustic phonon peaks in 6H–SiC, the origin was determined in the acoustic branch of the second-order Raman scattering as supported by theoretical calculations reported in literature.[6,27] The TO peak (801 cm^{-1}) was more intense in the perpendicular orientation, whereas the LO peak (970 cm^{-1}) was stronger in the parallel alignment.

Undoped 4H–SiC

Raman spectra were obtained for undoped 4H–SiC with the incident laser oriented parallel (Fig. 2(a)) and perpendicular (Fig. 2(b)) to the c-axis. Both spectra were recorded at about 10 °C sample temperature. The peaks at 200 cm^{-1} (split FTA, E_2) were more intense in the parallel than in the perpendicular orientation. The peak at 268 cm^{-1} (FTA, E_2), however, was more intense in perpendicular orientation. The FLA peak (610 cm^{-1}), FTO peak (780 cm^{-1}) and the LO peak (970 cm^{-1}) showed no dependency on orientation.

Nitrogen-doped 4H–SiC

The Raman spectrum in Fig. 3(a) was obtained for the nitrogen-doped 4H–SiC sample (approx. 5 × 10^{18} cm^{-3}) with the incident laser oriented parallel to the c-axis. Figure 3(b) illustrates the results for the same sample in perpendicular orientation (both measurements at about 10 °C). Here, the FTA peak at 200 cm^{-1} was similar in both spectra. The peak at 269 cm^{-1} (FTA, E_1) was stronger in perpendicular orientation. The folded TO peak (FTO, E_2) at 781 cm^{-1} was split into a doublet with 777 cm^{-1} and 783 cm^{-1} in perpendicular orientation. The undoped 4H–SiC sample did not show any splitting. In the perpendicular geometry, an additional peak appeared at 839 cm^{-1} (FLO, A_1) while the LOPC line at 999 cm^{-1} vanished. The FLO peak at 841 cm^{-1} – only present in perpendicular geometry – was not affected by doping. The FLO peak showed the same appearance for the doped and undoped SiC samples.

Table 2. Slopes of the Raman peaks shift in the temperature range 3–112 °C for SiC[a]

Undoped 4H parallel		Doped 4H parallel		Undoped 6H parallel	
Peak	Slope (cm^{-1}/K)	Peak	Slope (cm^{-1}/K)	Peak	Slope (cm^{-1}/K)
207.2	−0.0034	208.1	−0.0023	154.2	−0.0012
268.1	−0.0018	269.5	−0.0032	270.3	−0.0057
612.5	−0.0086	614.1	−0.0107	508.6	−0.0057
797.4	−0.0164	781.1	−0.0166	517.7	−0.0048
799.4	−0.0164	801.2	−0.0155	771.2	−0.0151
968.3	−0.0182	999.2	+0.0148	792.8	−0.0160
				970.7	−0.0207

Undoped 4H perpendicular		Doped 4H perpendicular		Undoped 6H perpendicular	
Peak	Slope (cm^{-1}/K)	Peak	Slope (cm^{-1}/K)	Peak	Slope (cm^{-1}/K)
268.1	−7.45E-4	204.7	−0.0038	153.9	−0.0030
612.2	−0.0056	266.6	−0.0070	244.3	−0.0011
785.4	−0.0157	610.6	−0.0111	270.3	−0.0044
800.3	−0.0144	777.7	−0.0144	509.3	−0.0058
840.9	−0.0131	798.9	−0.0146	516.8	−0.0056
974.6	−0.0157	798.9	−0.0164	771.2	−0.0152
		839.3	−0.0136	792.5	−0.0174
				801.6	−0.0158
				976.7	−0.0120

[a] The slopes are given for peaks measured in a parallel and perpendicular orientation of the incident laser to the c-axis. The numerical values can be used to calibrate spectroscopic temperature measurements.

Temperature shift

Using Eqn (1), the shift of a Raman peak center can be reproduced. For each peak in the spectra, the fitting constants C and D have to be calculated. C and D increase with higher wavenumbers of the Raman peaks. Parameter D can be neglected for temperatures below 500 °C, where its contribution to the temperature shift is negligible.[2] Thus, by expanding C to a linear function of ω_0 within this temperature range, the equation can be used for all Raman peaks. The equation describes the characteristics of the temperature shift of all peaks of the spectra very well.

Figure 4 shows the absolute shift $\Delta\omega = \omega - \omega_0$ of the peak centers induced by temperature for all samples and orientations. The peak centers showed a strictly linear dependence on temperature. From the Raman spectra recorded for increasing sample temperature, we determined a shift of all phonon peaks to smaller values. An exception to this dependence is the A_1(LOPC) line in the nitrogen-doped 4H–SiC sample, which was shifted to higher wavenumbers. The shift of the Raman lines depended linearly on the actual peak center position. Thus, a linear fit was performed on these data and the resulting slopes are shown in Table 2. For comparison, the slope of the linear temperature shift of the first-order Raman peak in silicon (−0.025 cm^{-1}/K at 521 cm^{-1}) is even stronger than the maximum slope measured for SiC (−0.021 cm^{-1}/K at 970 cm^{-1}).

It is striking that for the 6H–SiC sample, the slope values of the 509 and 517 cm^{-1} peaks are also located on the same straight line as all other coefficients for SiC samples (Fig. 5). Therefore, an origin from Si–Si bonds, appearing at 521 cm^{-1} for silicon, can be excluded. In the case of Si–Si bonds, the slope values should be far off this straight line (−0.025 cm^{-1}/K at 521 cm^{-1}).

The spectral position and line-shape of the A_1(LOPC) mode are known to be determined by the concentration and mobility of free carriers in doped SiC.[6,11–13,28] Raman measurements at constant temperatures for differently doped samples revealed that with increasing free carrier concentration, the LOPC peak position shifts to higher wavenumbers.[6,14] In our measurements, we detected that the peak center of the LOPC mode in doped 4H–SiC shifted linearly to higher values with increasing temperature ($\Delta\omega = 1.6$ cm^{-1} for $\Delta T = 110$ K). The asymmetry of the peak shape did not vary significantly with temperature. For p- and n-type 4H–SiC, an increase of the free electron concentration for increasing temperature has been reported.[14,16] This suggests that in our measurements, the increasing temperature can lead to a higher concentration of free carriers. Thus, the increase of the concentration shifts the LOPC mode to higher

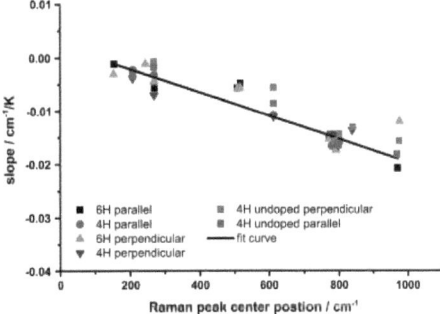

Figure 5. Slopes of the temperature shift for all peaks. This figure is available in colour online at www.interscience.wiley.com/journal/jrs.

wavenumbers, overcompensating the temperature-induced shift to lower wavenumbers.

Focusing on the position of the Raman peak center (ω_0) within the spectra and the corresponding slope of the temperature shift, the slope is increasing with higher peak center position. This curve shows a nearly linear behavior as well (illustrated in Fig. 5). The energy transfer during the inelastic scattering increases. From our data, we derive the coefficient for the temperature-induced line-shift: $C(\omega_0) = -2.188 \times 10^{-5} \omega_0 + 0.002$ cm^{-1}.

Conclusion

The peak positions of the Raman lines in SiC change with varying temperature. We have investigated their temperature dependence in the range from 3 to 112 °C. We observed a linear relation between line-shift and temperature. The slope of the temperature shift increases with increased energy transfer due to inelastic scattering showing a linear dependency with temperature in the temperature range of about $\Delta T = 110$ K. Although the peak center position is typically shifted to lower wavenumbers with increasing temperature, the nitrogen-doped plasmon-coupled longitudinal optical phonon mode (LOPC) showed a contrary behavior, moving to higher values.

For accurate stress and temperature measurements in SiC devices, the linearity of the temperature shift within the typical operation temperatures simplifies spectroscopic temperature measurements. We found that, in contrast to silicon, the effect of local heating by the laser at 532 nm wavelength can be neglected in confocal Raman spectroscopy due to the small absorption coefficient of SiC. Our results provide calibration data for spectral temperature measurements in SiC that simplify the spectroscopic characterization of electrical properties, mechanical stress and local temperatures.

Acknowledgements

We thank the Deutsche Forschungsgemeinschaft (DFG) cluster of excellence 'Nanosystems Initiative Munich' for financial support.

References

[1] T. Beechem, S. Graham, S. P. Kearney, L. M. Phinney, J. R. Serrano, *Rev. Sci. Instrum.* **2007**, *78*.
[2] H. Harima, T. Hosoda, S. Nakashima, *J. Electron. Mater.* **1999**, *28*, 141.
[3] K. Dombrowski, *Micro-Raman investigation of mechanical stress in Si device structures and phonons in SiGe*, Dissertation thesis, TU-Cottbus (Cottbus), **2000**.
[4] D. W. Feldman, J. H. Parker, W. J. Choyke, L. Patrick, *Phys. Rev.* **1968**, *173*, 787.
[5] D. W. Feldman, J. H. Parker, W. J. Choyke, L. Patrick, *Phys. Rev.* **1968**, *170*, 698.
[6] J. C. Burton, L. Sun, M. Pophristic, S. J. Lukacs, F. H. Long, Z. C. Feng, I. T. Ferguson, *J. Appl. Phys.* **1998**, *84*, 6268.
[7] S. Nakashima, H. Harima, *Phys. Stat. Sol. A* **1997**, *162*, 39.
[8] J. C. Burton, L. Sun, F. H. Long, Z. C. Feng, I. T. Ferguson, *Phys. Rev. B* **1999**, *59*, 7282.
[9] M. V. Klein, P. J. Colwell, B. N. Ganguly, *Phys. Rev. B* **1972**, *6*, 2380.
[10] M. Roschke, F. Schwierz, *IEEE T. Electron. Dev.* **2001**, *48*, 1442.
[11] H. Harima, *Microelectron. Eng.* **2006**, *83*, 126.
[12] S. Nakashima, H. Harima, *J. Appl. Phys.* **2004**, *95*, 3541.
[13] S. Nakashima, T. Mitani, J. Senzaki, H. Okumura, T. Yamamoto, *J. Appl. Phys.* **2005**, *97*.
[14] M. Laube, F. Schmid, G. Pensl, G. Wagner, M. Linnarsson, M. Maier, *J. Appl. Phys.* **2002**, *92*, 549.
[15] A. Compaan, H. J. Trodahl, *Phys. Rev. B* **1984**, *29*, 793.
[16] D. Olego, M. Cardona, *Phys. Rev. B* **1982**, *25*, 3889.
[17] D. G. Mead, G. R. Wilkinson, *J. Raman Spectrosc.* **1977**, *6*, 123.
[18] Z. Li, R. C. Bradt, *J. Mater. Sci.* **1986**, *21*, 4366.
[19] Z. Li, R. C. Bradt, *J. Am. Ceram. Soc.* **1987**, *70*, 445.
[20] C. Ulrich, A. Debernardi, E. Anastassakis, K. Syassen, M. Cardona, *Phys. Stat. Sol. B* **1999**, *211*, 293.
[21] J. Menendez, M. Cardona, *Phys. Rev. B* **1984**, *29*, 2051.
[22] T. R. Hart, R. L. Aggarwal, B. Lax, *Phys. Rev. B* **1970**, *1*, 638.
[23] R. Tsu, J. G. Hernandez, *Appl. Phys. Lett.* **1982**, *41*, 1016.
[24] S. Kouteva-Arguirova, T. Arguirov, D. Wolfframm, J. Reif, *J. Appl. Phys.* **2003**, *94*, 4946.
[25] A. Debernardi, C. Ulrich, K. Syassen, M. Cardona, *Phys. Rev. B* **1999**, *59*, 6774.
[26] M. Bauer, A. M. Gigler, C. Richter, R. W. Stark, *Microelectron. Eng.* **2008**, *85*, 1443.
[27] W. Windl, K. Karch, P. Pavone, O. Schutt, D. Strauch, W. H. Weber, K. C. Hass, L. Rimai, *Phys. Rev. B* **1994**, *49*, 8764.
[28] H. Harima, S. Nakashima, T. Uemura, *J. Appl. Phys.* **1995**, *78*, 1996.

7.3 Nanoscale residual stress-field mapping around nanoindents in SiC by IR s-SNOM and confocal Raman microscopy

Alexander M. Gigler, Andreas J. Huber, Michael Bauer, Alexander Ziegler, Rainer Hillenbrand and Robert W. Stark

Optics Express, Vol. 17, Issue 25, pp. 22351-22357

Permanent weblink:
http://dx.doi.org/10.1364/OE.17.022351

Nanoscale residual stress-field mapping around nanoindents in SiC by IR s-SNOM and confocal Raman microscopy

Alexander M. Gigler[1,3,*], Andreas J. Huber[2,3,4], Michael Bauer[1,3], Alexander Ziegler[4], Rainer Hillenbrand[2,3,5], and Robert W. Stark[1,3]

[1] *Department of Earth and Environmental Sciences, Ludwig-Maximilians-Universität München, Theresienstr. 41, 80333 Munich, Germany*
[2] *Nanooptics Laboratory, CIC nanoGUNE Consolider, 20018 Donostia – San Sebastian, Spain*
[3] *Center for NanoScience (CeNS), Schellingstr. 4, 80799 Munich, Germany*
[4] *Molekulare Strukturbiologie, Max-Planck-Insitut für Biochemie, 82152 Martinsried, Germany*
[5] *IKERBASQUE, Basque Foundation for Science*
*gigler@lmu.de

Abstract: We map a nanoindent in a silicon carbide (SiC) crystal by infrared (IR) scattering-type scanning near-field optical microscopy (s-SNOM) and confocal Raman microscopy and interpret the resulting images in terms of local residual stress-fields. By comparing near-field IR and confocal Raman images, we find that the stress-induced shifts of the longitudinal optical phonon-frequencies (LO) and the related shift of the phonon-polariton near-field resonance give rise to Raman and s-SNOM image contrasts, respectively. We apply single-frequency IR s-SNOM for nanoscale resolved imaging of local stress-fields and confocal Raman microscopy to obtain the complete spectral information about stress-induced shifts of the phonon frequencies at diffraction limited spatial resolution. The spatial extension of the local stress-field around the nanoindent agrees well between both techniques. Our results demonstrate that both methods ideally complement each other, allowing for the detailed analysis of stress-fields at e.g. material and grain boundaries, in Micro-Electro-Mechanical-Systems (MEMS), or in engineered nanostructures.

©2009 Optical Society of America

OCIS codes: (180.4243) Near-field microscopy; (180.5655) Raman microscopy; (240.3695) Linear and nonlinear light scattering from surfaces; (240.5420) Polaritons; (240.6680) Surface plasmons; (300.6330) Spectroscopy, inelastic scattering including Raman

References and links

1. D. Peter, M. Dalmer, H. Kruwinus, A. Lechner, L. Archer, E. Gaulhofer, A. M. Gigler, R. W. Stark, and W. Bensch, "Measurement of the Mechanical Stability of Semiconductor Line Structures in Relevant Media," ECS Trans. **16**, 13–21 (2009).
2. I. Neizvestnyi, and V. Gridchin, "The use of stressed silicon in MOS transistors and CMOS structures," Russ. Microelectron. **38**(2), 71–86 (2009).
3. J. Baliga, *Silicon Carbide Power Devices* (World Scientific, Singapore, 2005).
4. H. Harima, T. Hosoda, and S. Nakashima, "Temperature measurement in a silicon carbide light emitting diode by Raman scattering," J. Electron. Mater. **28**(3), 141–143 (1999).
5. A. J. Wilkinson, G. Meaden, and D. J. Dingley, "High-resolution elastic strain measurement from electron backscatter diffraction patterns: new levels of sensitivity," Ultramicroscopy **106**(4-5), 307–313 (2006).
6. A. J. Wilkinson, G. Meaden, and D. J. Dingley, "Mapping strains at the nanoscale using electron back scatter diffraction," Superlattices Microstruct. **45**(4-5), 285–294 (2009).
7. M. Bauer, A. M. Gigler, A. J. Huber, R. Hillenbrand, and R. W. Stark, "Temperature depending Raman line-shift in silicon carbide," J. Raman Spectrosc. (to be published), doi:10.1002/jrs.2334.
8. M. Bauer, A. M. Gigler, C. Richter, and R. W. Stark, "Visualizing stress in silicon micro cantilevers using scanning confocal Raman spectroscopy," Microelectron. Eng. **85**(5-6), 1443–1446 (2008).
9. T. Beechem, S. Graham, S. P. Kearney, L. M. Phinney, and J. R. Serrano, "Invited Article: Simultaneous mapping of temperature and stress in microdevices using micro-Raman spectroscopy," Rev. Sci. Instrum. **78**(6), 061301 (2007).

10. D. Olego, M. Cardona, and P. Vogl, "Pressure-Dependence of the Optical Phonons and Transverse Effective Charge in 3C-SiC," Phys. Rev. B **25**(6), 3878–3888 (1982).
11. H. F. Poulsen, J. A. Wert, J. Neuefeind, V. Honkimäki, and M. Daymond, "Measuring strain distributions in amorphous materials," Nat. Mater. **4**(1), 33–36 (2005).
12. A. Debernardi, C. Ulrich, K. Syassen, and M. Cardona, "Raman linewidths of optical phonons in 3C-SiC under pressure: First-principles calculations and experimental results," Phys. Rev. B **59**(10), 6774–6783 (1999).
13. T. B. Wei, Q. Hu, R. F. Duan, J. X. Wang, Y. P. Zeng, J. M. Li, Y. Yang, and Y. L. Liu, "Mechanical Deformation Behavior of Nonpolar GaN Thick Films by Berkovich Nanoindentation," Nanoscale Res. Lett. **4**(7), 753–757 (2009).
14. J. C. Burton, L. Sun, M. Pophristic, S. J. Lukacs, F. H. Long, Z. C. Feng, and I. T. Ferguson, "Spatial characterization of doped SiC wafers by Raman spectroscopy," J. Appl. Phys. **84**(11), 6268–6273 (1998).
15. S. Nakashima, and H. Harima, "Raman investigation of SiC polytypes," Phys. Status Solidi A **162**(1), 39–64 (1997).
16. K. Mizoguchi, and S. Nakashima, "Determination of Crystallographic Orientations in Silicon Films by Raman-Microprobe Polarization Measurements," J. Appl. Phys. **65**(7), 2583–2590 (1989).
17. I. DeWolf, "Micro-Raman spectroscopy to study local mechanical stress in silicon integrated circuits," Semicond. Sci. Technol. **11**(2), 139–154 (1996).
18. S. M. Hu, "Stress-Related Problems in Silicon Technology," J. Appl. Phys. **70**(6), R53–R80 (1991).
19. B. V. Kamenev, H. Grebel, L. Tsybeskov, T. I. Kamins, R. S. Williams, J. M. Baribeau, and D. J. Lockwood, "Polarized Raman scattering and localized embedded strain in self-organized Si/Ge nanostructures," Appl. Phys. Lett. **83**(24), 5035–5037 (2003).
20. J. Liu, and Y. K. Vohra, "Raman modes of 6H polytype of silicon carbide to ultrahigh pressures: A comparison with silicon and diamond," Phys. Rev. Lett. **72**(26), 4105–4108 (1994).
21. J. Liu, and Y. K. Vohra, "Raman modes of 6H polytype of silicon carbide to ultrahigh pressures - Reply," Phys. Rev. Lett. **77**, 1661 (1996).
22. L. G. Cançado, A. Hartschuh, and L. Novotny, "Tip-enhanced Raman spectroscopy of carbon nanotubes," J. Raman Spectrosc. **40**(10), 1420–1426 (2009).
23. A. Hartschuh, E. J. Sánchez, X. S. Xie, and L. Novotny, "High-resolution near-field Raman microscopy of single-walled carbon nanotubes," Phys. Rev. Lett. **90**(9), 095503 (2003).
24. D. Cialla, T. Deckert-Gaudig, C. Budich, M. Laue, R. Moller, D. Naumann, V. Deckert, and J. Popp, "Raman to the limit: tip-enhanced Raman spectroscopic investigations of a single tobacco mosaic virus," J. Raman Spectrosc. **40**(3), 240–243 (2009).
25. T. Deckert-Gaudig, F. Erver, and V. Deckert, "Transparent silver microcrystals: synthesis and application for nanoscale analysis," Langmuir **25**(11), 6032–6034 (2009).
26. T. Deckert-Gaudig, E. Bailo, and V. Deckert, "Perspectives for spatially resolved molecular spectroscopy - Raman on the nanometer scale," J. Biophoton. **1**(5), 377–389 (2008).
27. N. Hayazawa, M. Motohashi, Y. Saito, H. Ishitobi, A. Ono, T. Ichimura, P. Verma, and S. Kawata, "Visualization of localized strain of a crystalline thin layer at the nanoscale by tip-enhanced Raman spectroscopy and microscopy," J. Raman Spectrosc. **38**(6), 684–696 (2007).
28. A. Tarun, N. Hayazawa, M. Motohashi, and S. Kawata, "Highly efficient tip-enhanced Raman spectroscopy and microscopy of strained silicon," Rev. Sci. Instrum. **79**(1), 013706 (2008).
29. F. Keilmann, and R. Hillenbrand, "Near-field microscopy by elastic light scattering from a tip," Philos. Trans. R. Soc. Lond. A **362**(1817), 787–805 (2004).
30. R. Hillenbrand, T. Taubner, and F. Keilmann, "Phonon-enhanced light matter interaction at the nanometre scale," Nature **418**(6894), 159–162 (2002).
31. S. C. Kehr, M. Cebula, O. Mieth, T. Härtling, J. Seidel, S. Grafström, L. M. Eng, S. Winnerl, D. Stehr, and M. Helm, "Anisotropy contrast in phonon-enhanced apertureless near-field microscopy using a free-electron laser," Phys. Rev. Lett. **100**(25), 256403 (2008).
32. A. Huber, N. Ocelic, T. Taubner, and R. Hillenbrand, "Nanoscale resolved infrared probing of crystal structure and of plasmon-phonon coupling," Nano Lett. **6**(4), 774–778 (2006).
33. N. Ocelic, and R. Hillenbrand, "Subwavelength-scale tailoring of surface phonon polaritons by focused ion-beam implantation," Nat. Mater. **3**(9), 606–609 (2004).
34. A. J. Huber, A. Ziegler, T. Köck, and R. Hillenbrand, "Infrared nanoscopy of strained semiconductors," Nat. Nanotechnol. **4**(3), 153–157 (2009).
35. N. Ocelic, A. Huber, and R. Hillenbrand, "Pseudoheterodyne detection for background-free near-field spectroscopy," Appl. Phys. Lett. **89**(10), 101124 (2006).
36. U. Schmidt, W. Ibach, J. Muller, K. Weishaupt, and O. Hollricher, "Raman spectral imaging - A nondestructive, high resolution analysis technique for local stress measurements in silicon," Vib. Spectrosc. **42**(1), 93–97 (2006).
37. T. Wermelinger, C. Borgia, C. Solenthaler, and R. Spolenak, "3-D Raman spectroscopy measurements of the symmetry of residual stress fields in plastically deformed sapphire crystals," Acta Mater. **55**(14), 4657–4665 (2007).
38. G. D. Quinn, and R. C. Bradt, "On the Vickers indentation fracture toughness test," J. Am. Ceram. Soc. **90**(3), 673–680 (2007).
39. M. Becker, H. Scheel, S. Christiansen, and H. P. Strunk, "Grain orientation, texture, and internal stress optically evaluated by micro-Raman spectroscopy," J. Appl. Phys. **101**(6), 063531 (2007).

1. Introduction

Mechanical and residual strain in semiconductor materials play an important role in the design of devices. For example, strain can give rise to failure, in particular due to the collapse of high aspect-ratio features [1]. Controlled strain engineering, however, allows affecting free carrier mobility in semiconductors and may enable manipulation and customization of novel electronic devices at the nano-scale [2]. One of the more robust semiconductor materials is silicon carbide (SiC). It is used in demanding applications under elevated temperatures, high voltage, and for optical applications in the ultra violet region [3,4]. Thus, local strain measurements are necessary for monitoring and optimization of fabrication processes. For this purpose, it is desirable to use non-invasive and fast techniques for strain mapping that require only minimum sample preparation. Precise local strain measurements can be, for example, achieved by electron backscatter diffraction [5,6]. For routine analysis, however, optical methods are desirable because they require less preparation. As the frequency of phonon modes is changing under external influences such as mechanical stress and strain [7–13], both infrared near-field microscopy and Raman spectroscopy are well suited techniques.

Raman scattering in SiC occurs on longitudinal and transversal acoustic (LA, TA) and on optical (LO, TO) phonon modes. Scattering at folded modes can occur in higher order SiC polytypes due to the increased size of the unit cell [14,15]. The orientation of the crystal axes with respect to the actual polarization of the incident laser, can cause anisotropic Raman scattering signals [16]. Together with the strain tensor, the effect of polarization has been investigated for silicon [17,18] and silicon-germanium nanostructures [19]. Raman spectroscopy is highly sensitive to variations in phonon frequencies, i.e. the line shift. These variations can be caused by modifications of the equilibrium distance between atoms due to either mechanical stress or changing temperature [9]. Because the peak width is mainly depending on the sample temperature, whereas the peak position is affected by both stress and temperature, these effects can be separated. However, high mechanical stress [20,21] as well as defects may also influence the line width [17]. In general, not only external influences but also intrinsic effects of the measurement process, such as the strongly focused laser, can cause heating of the sample and – in turn – line shifts. In the case of SiC, however, such a local heating effect can be excluded for moderate laser intensities, because the absorption coefficient of SiC (1.59 per cm at 532 nm) is very small and the penetration depth is about 2 mm [7]. Hence, confocal Raman spectroscopy enables a direct measurement of the local distribution of mechanical stress and strain in SiC samples with a lateral resolution determined by the confocal spot size.

Confocal techniques are limited by diffraction. The diffraction limit can be overcome by optical near-field techniques. For example, tip-enhanced Raman spectroscopy can provide a sub-wavelength lateral resolution [22–28]. Another highly resolving spectroscopic technique is scattering-type scanning near-field optical microscopy (s-SNOM) [29]. At infrared frequencies, s-SNOM can probe the structural properties of infrared-active materials such as SiC, SiN, SiO, GaN, or GaAs with nanometer resolution. In s-SNOM the metalized tip of an Atomic Force Microscope (AFM) is illuminated by laser light. The concentrated optical fields at the tip apex can excite localized surface phonon-polaritons (optical lattice vibrations), yielding a near-field resonance at a material specific frequency ω_{sp} close to the LO phonon frequency [30,31]. As the phonon frequencies strongly depend on the crystal lattice, spectroscopic recording of the backscattered infrared radiation allows for nano-scale mapping of polytypism [32], crystal quality [33], or anisotropy [31]. The surface phonon-polariton near-field resonance of SiC also varies with compressive or tensile strain, which enables imaging of strain-fields [34].

Here, we use confocal Raman microscopy to map the stress-fields introduced in a 6H-SiC sample by means of nanoindentation. We focus on the stress-induced line shifts of the TO and LO phonon modes of the sample. In addition we compare our results to s-SNOM measurements of the identical nanoindent. Our study shows that confocal Raman microscopy and s-SNOM probe the direct effects of stress on the phonon spectrum of the sample.

The differences in axial resolution and, thus, image contrast formation are discussed in the following. Furthermore, we demonstrate the high-resolution capabilities of s-SNOM, revealing nanocracks at a resolution far beyond the diffraction limit of a confocal Raman microscope. The combination of both analytic techniques proves highly effective for the assessment of local stress distributions with full spectral information as well as high resolution data at a single frequency.

2. Materials and methods

We investigated the hexagonal polytype of silicon carbide (6H-SiC). The sample was a single crystal of 6H-SiC with (0001) orientation polished to epitaxy-ready grade. Due to the hexagonal symmetry of the 6H-SiC crystal and the (0001) orientation of the crystallographic axis, the s-SNOM contrast is independent of the illumination direction for the given experimental setup. To generate local stress fields at the surface of the SiC crystal, a NanoTest 600 (Micro Materials Ltd.) device equipped with a pyramidal indenter (Berkovich) was used. The loading and unloading rate for the indentation was 0.5 mN/s, while the force was held constant at maximum load for 5 s. The nanoindentations were created with a maximum load of 50 mN resulting in 130 nm deep elasto-plastic deformations (triangular depressions). The identical indent was examined by both confocal Raman microscopy and s-SNOM.

Raman spectra were recorded by a confocal Raman microscope (alpha300 R; WITec GmbH, Ulm, Germany). The system was equipped with a SHG Nd:YAG laser (532 nm, 22.5 mW) and a lens-based spectrometer. Elastically scattered photons were rejected by a sharp edge filter. With a 1800 mm^{-1} grating, the spectral resolution was 1.17 cm^{-1} per CCD-pixel. The Raman peaks were fitted with a Lorentzian profile. Spectra were recorded at each image pixel with an integration time of 0.4 s. Due to the band-gap of 3.0 eV for 6H-SiC compared to the photon energy of 2.33 eV, the scattering process was non-resonant. A 100 × microscope objective (working distance 0.26 mm, NA 0.90) was used for the measurements. Due to the holographic beam splitter, the incident light was polarized along the x-axis of the scan. However, our measurements can be regarded as "unpolarized", because no analyzer was used. Thus, a stress magnitude is measured rather than the exact components of the strain tensor. The diffraction limited focus resulted in a lateral resolution of about 400 nm. A multimode optical fiber guided the signal to the spectrometer. The 50-µm-core of this fiber was used as confocal pinhole leading to an axial focal depth of about 1 µm.

The scattering-type scanning near-field optical microscope (s-SNOM) used for this work is based on a custom-built atomic force microscope (AFM). The technical details have been reported elsewhere [30,32–35]. In this setup, the metalized tip of an AFM cantilever is illuminated by a frequency-tunable CO_2 laser providing wavelengths between $\omega_{IR} = 887$ cm^{-1} and $\omega_{IR} = 1085$ cm^{-1} ($\lambda_{IR} = 9.21$ µm – 11.27 µm). The backscattered light was recorded simultaneously with topography providing ultra-high resolution infrared near-field images.

3. Result and discussion

The peaks in the Raman spectra of unstressed SiC were identified and assigned to the phonon lines of 6H-SiC according to literature [15]: 234.6 cm^{-1} (folded TA-planar), 240.0 cm^{-1} (FTA-planar), 505.5 cm^{-1} (FLA-axial), 787.9 cm^{-1} (FTO-planar), 798.0 cm^{-1} (FTO-planar), 889.3 cm^{-1} (TO-planar), and 971.5 cm^{-1} (LO-axial). In the following, we discuss the optical phonon-lines ($\omega_{TO} = 889.3$ cm^{-1} and $\omega_{LO} = 971.5$ cm^{-1}). Figure 1 shows the Raman spectrum of 6H-SiC in the region of optical phonons ($\omega = 775$ cm^{-1} – 990 cm^{-1}). The inset shows the change in the spectra for different loading conditions taken from different regions around the indent. The green curve indicates the position and curve shape of the TO and LO peaks obtained for the unstressed material. With respect to the excitation wavelength, the peaks are shifted to lower wavenumbers for tensile stress, while for compressive stress they are shifted to higher values. In analogy to Liu and Vohra's work [20,21], we can estimate a maximal mechanical stress around the indent of about ± 1 GPa from the line-shift of about ± 2 cm^{-1}.

This conclusion is supported by the increasing separation of the TO and LO center frequencies with increasing pressure, yielding a similar value for the stress magnitude.

Note that the precise calculation of the local stress and strain from the shift of the phonon frequencies depends on crystal orientation and polarization of the incident light. As our measurements were unpolarized, the local stress obtained here corresponds to a scalar average rather than exact components of the strain tensor.

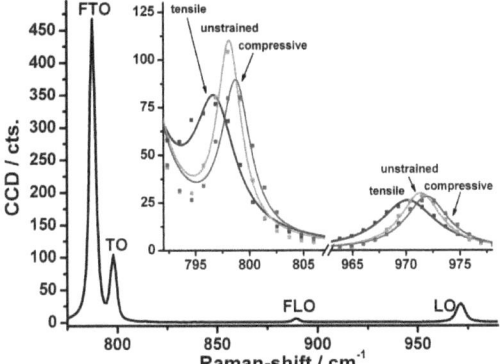

Fig. 1. (color online): Typical Raman spectrum of unstressed 6H-SiC of the most prominent phonon lines at 889.3 cm^{-1} (TO-planar) and 971.5 cm^{-1} (LO-axial). The inset shows the TO and LO peaks under tensile (blue), neutral (green), and compressive (red) loading causing corresponding line-shifts and line-broadening (symbols: experiment; lines: Lorentzian fit).

We examined the identical indent in both Raman and s-SNOM measurements (Fig. 2). All images in Fig. 2 show a 15 µm × 15 µm scan field. A map of the elastically scattered intensity (Rayleigh scattering) recorded at diffraction limited spatial resolution by the confocal Raman microscope is shown in Fig. 2(a). Regions of interest were labeled A-D. In order to obtain stress-distribution images from the Raman spectral data, the phonon lines were fitted by Lorentzian profiles and the center frequencies (Figs. 2(b) and 2(c)) of the peaks were determined for each pixel. The center frequencies were shifted to higher wavenumbers where compressive stress was present (close to the edges of the indent) and to lower wavenumbers where tensile stress occurred (corners of the indent) [36]. The maximum shift amounts to about ± 2 cm^{-1} for the TO and LO phonons. As seen in Figs. 2(b) and 2(c), compressive stress was induced in regions A and C, while in the regions B and D near the indent mainly tensile stress occurred. In Figs. 2(b) and 2(c) the stress field around the indent is not perfectly symmetric. The asymmetry of the stress pattern, i.e. that we find tensile stress instead of compression at position D, we explain by a slight misalignment of the indenter, i.e. the indentation was not performed perfectly perpendicular to the sample surface. In addition, due to the axial focus of approximately 1 µm, not only the topmost stressed surface of the SiC sample was probed, but also stress distributions beneath the surface of the crystal can contribute to the spectra. Thus, an effective stress, i.e. an average stress of the sampled depth, was measured rather than one highly localized at the surface.

Figures 2(d) - 2(f) show s-SNOM images of the indent in 6H-SiC. The triangular depression of 130 nm depth is located in the center of the topography image (Fig. 2(d)). The topography image does not reveal any stress-related features. In the infrared near-field images (Figs. 2(e) and 2(f)), strong contrasts can be observed in the proximity of the indent. The near-field contrast reveals the characteristic pattern of residual stress-fields as detected in the confocal Raman images (see Figs. 2(b) and 2(c)), yet at much higher spatial resolution. In the near-field image recorded at ω_{IR} = 924 cm^{-1}, regions near the edges of the indent appear in dark and near the corners in bright contrast as compared to the undisturbed surface region.

This contrast sequence is reversed for $\omega_{IR} = 944$ cm^{-1}. The inversion of the image contrast is due to the fact that the s-SNOM images are recorded at single frequencies, slightly below and above the resonance frequency of the surface phonon-polariton near-field resonance [32].

The amplitude contrast of the s-SNOM images of the indent can be assigned to stress-induced shifts of this surface phonon-polariton near-field resonance. The s-SNOM images reveal compressive stress at the edges of the indent and tensile stress at its corners [34].

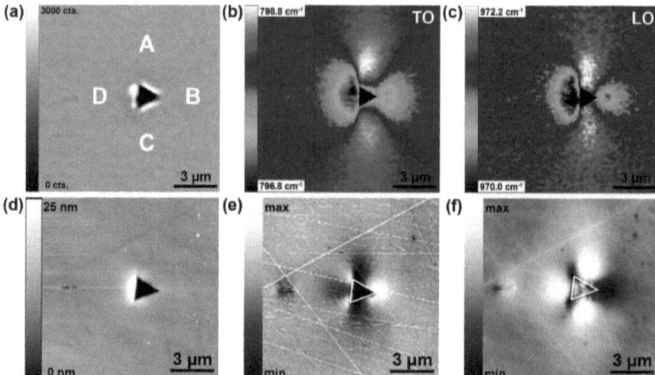

Fig. 2. (color online): Confocal Raman microscopy and s-SNOM images of a locally stressed SiC sample. (a) Rayleigh intensity map. Four regions of interest are labeled A-D, (b, c) maps of the spectral position of the fitted TO and LO phonon lines as obtained by fitting a Lorentzian. Black triangles mark the contour of the indent. For comparison, s-SNOM amplitude maps of the identical indent are given in (d-f). (d) Shows the topography of the indent, (e) s-SNOM amplitude detected for $\omega_{IR} = 924$ cm^{-1} and (f) $\omega_{IR} = 944$ cm^{-1} laser frequencies reveal the characteristic amplitude contrasts around the indent. Note that in the near-field optical images the scratches exhibit a pronounced contrast, which can be explained by crystal damage caused by the mechanical polishing process [33].

Comparison of the confocal Raman maps and the IR near-field images of the indent reveals that both methods exhibit stress-induced contrasts around the nanoindent caused by the indentation process. The spatial extension and geometry of the lobes around the indent agree well between the confocal Raman and the IR s-SNOM images. A deviation between near-field s-SNOM and confocal Raman maps was found in Figs. 2(b), 2(c), 2(e), and 2(f) at the position D of the indent. In the s-SNOM images, highly localized tensile stress-fields were identified at the corners in region D of the indent as well as a larger compressively stressed sample area at position D. The confocal Raman images, on the other hand, revealed only tensile stress for this specific sample area. In order to understand the differences in image contrast formation at region D, two effects have to be considered. First, while s-SNOM probes the very surface of the sample, confocal Raman scattering occurs within surface-near layers up to several 100 nm beneath the surface. Due to the averaging over the entire confocal volume, the signal of Raman scattering was dominated by the differently stressed material deeper inside the 6H-SiC crystal. Such an averaging effect has also been observed by Wermelinger et al. for a Vickers indentation in GaN [37]. Second, the confocal Raman microscopy images reveal how stress affects the in-plane (parallel to sample surface) TO and LO phonon frequencies. In s-SNOM, the probing infrared field at the tip apex is largely depolarized, thus probing stress-induced changes of the phonon polaritons both parallel and perpendicular to the sample surface. Because of the slightly misaligned indenter, we can assume that the stress distributions perpendicular and parallel to the sample surface differ from each other, yielding the differences in the stress patterns observed by confocal Raman and s-SNOM images.

The potential of infrared s-SNOM for the mapping of nano-scale stress-fields is demonstrated by imaging a corner of the indent at higher resolution as shown in the topography image in Fig. 3(a). The corresponding amplitude image (Fig. 3(b)) recorded at ω_{IR} = 935 cm^{-1} clearly reveals a dark line emanating from the corner of the indent (corresponding to position B in Fig. 2(a)) with a circular shaped contrast at its end.

Congruent results of finite element calculations and macroscopic investigations of similar configurations confirm, that at such critical edge locations of indents where tensile stress occurs, formation of such cracks is preferential [38]. The bulb-shaped feature observed at the end of the dark line, exhibits an amplitude contrast of a tensile stressed sample area, which is characteristic for the stress-field at crack tips [36,39]. The dark line, thus, corresponds to the signature of the nanocrack. Indeed, the topography at an even smaller scan range reveals also a small crack in the sample surface.

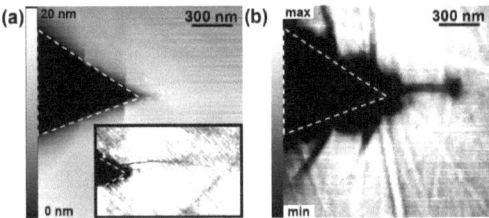

Fig. 3. (color online): s-SNOM hardware zoom into the right corner of the indent revealing a nanocrack emanating from the edge. (a) Topography and (b) s-SNOM amplitude at ω_{IR} = 935 cm^{-1}. The inset in (a) shows an additional hardware-zoom of the corner of the indent revealing a nanocrack in the topography data.

4. Conclusion

In summary we have shown that confocal Raman microscopy and s-SNOM imaging reveal a qualitative contrast of the stress-fields around nanoindentation sites. Both data sets correspond in their spatial extension. The resolution of the methods differs considerably, s-SNOM being approximately 10-20 times better. The shift of the phonon frequencies as probed by Raman scattering can be directly related to local stress similarly as it was detected for the near-field phonon-polariton resonance. The magnitude of the phonon line shift agrees well with the shift of the phonon-polariton near-field resonance. Thus, based on the near-field scattering intensity of a single selected wavelength, s-SNOM allows to resolve the smallest features of the indent down to nanocracks while diffraction limited confocal Raman spectroscopy gives the full spectral information of the changes in phonon dispersion. Scattering-SNOM is sensitive to the stress distribution near the surface, while far-field Raman scattering also occurs beneath the surface and can reveal changes in sub-surface stress distributions.

Acknowledgements

We thank the DFG cluster of excellence "Nanosystems Initiative Munich" (NIM) for financial support.

7.4 Exterior surface damage of calcium fluoride out coupling mirrors for DUV lasers

Michael Bauer, Martin Bischoff, Sigrid Jukresch, Thomas Hülsenbusch, Ansgar Matern, Andreas Görtler, Robert W. Stark and Ute Kaiser

Optics Express, Vol. 17, Issue 10, pp. 8253-8263

Permanent weblink:
http://dx.doi.org/10.1364/OE.17.008253

The supporting media file may be found under the address:
http://www.opticsinfobase.org/oe/viewmedia.cfm?URI=oe-17-10-8253-1

Exterior surface damage of calcium fluoride outcoupling mirrors for DUV lasers

Michael Bauer,[1,2,*] Martin Bischoff,[3] Sigrid Jukresch,[2] Thomas Hülsenbusch,[2] Ansgar Matern,[2] Andreas Görtler,[2] Robert W. Stark[1], Andrey Chuvilin[4] and Ute Kaiser[4]

[1]*Center for NanoScience (CeNS) and Department of Earth and Environmental Sciences, Ludwig-Maximilians-Universität München, Theresienstr. 41, 80333 Munich, Germany*
[2]*Coherent GmbH, Zielstattstraße 32, 81379 Munich, Germany*
[3]*Friedrich-Schiller-Universität, Institute of Applied Physics, Max-Wien-Platz 1, 07743 Jena, Germany*
[4]*Central Facility of Electron Microscopy, Electron Microscopy Group of Materials Science, University of Ulm, Albert-Einstein-Allee 11, 89069 Ulm, Germany*
[*]*email: michael.bauer@lrz.uni-muenchen.de*

Abstract: Damage of optical components due to laser irradiation reduces reliability and limits durability. Calcium fluoride (CaF_2) is commonly used for deep UV laser optics because it shows a very low tendency of color center formation as, compared to other UV-X optical materials. Here, we report on the exterior damage of CaF_2 UV-X optics due to radiation with high pulse-energy densities (80 mJ/cm^2) from an ArF laser. At such high energy densities, damage occurs on the external resonator side. The damage is generated by a partial alteration of the CaF_2 substrate to crystalline $CaCO_3$ (calcite). The decomposition of CaF_2 is mainly driven by photochemical processes in the presence of water vapor, which are induced by the UV-laser light and the elevated temperature within the beam profile.

©2009 Optical Society of America

OCIS codes: 140.3330 (Laser damage); 230.4040 (Mirrors); 140.2180 (Excimer lasers); 260.7200 (Ultraviolet, extreme).

References and links

1. J. H. Apfel, E. A. Enemark, D. Milam, W. L. Smith, and M. J. Weber, "Effects of barrier layers and surface smoothness on 150-ps, 1.064 μm laser damage of AR coatings on glass," in (NIST Spec. Pub. 5009, 1977), p. 255.
2. V. Liberman, M. Rothschild, J. H. C. Sedlacek, R. S. Uttaro, A. Grenville, A. K. Bates, and C. Van Peski, "Excimer-laser-induced degradation of fused silica and calcium fluoride for 193-nm lithographic applications," Opt. Lett. **24**, 58-60 (1999).
3. N. L. Boling, M. D. Crisp, and G. Dube, "Laser-Induced Surface Damage," Appl. Opt. **12**, 650-660 (1973).
4. T. Feigl, J. Heber, A. Gatto, and N. Kaiser, "Optics developments in the VUV - soft X-ray spectral region," Nucl. Instrum. Methods Phys. Res. A **483**, 351-356 (2002).
5. W. Arens, D. Ristau, J. Ullmann, C. Zaczek, R. Thielsch, N. Kaiser, A. Duparre, O. Apel, K. Mann, R., H. Lauth, H. Bernitzki, J. Ebert, S. Schippel, and H. Heyer, "Properties of fluoride DUV excimer laser optics: influence of the dielectric layers," Proc. SPIE **3902**, 250-259 (2000).
6. M. Marsi, A. Locatelli, M. Trovo, R. P. Walker, M. E. Couprie, D. Garzella, L. Nahon, D. Nutarelli, E. Renault, A. Gatto, N. Kaiser, L. Giannessi, S. Gunster, D. Ristau, M. W. Poole, and A. Taleb-Ibrahimi, "UV/VUV free electron laser oscillators and applications in materials science," Surf. Rev. Lett. **9**, 599-607 (2002).
7. D. Ristau, W. Arens, S. Bosch, A. Duparre, E. Masetti, D. Jacob, G. Kiriakidis, F. Peiro, E. Quesnel, and A. V. Tikhonravov, "UV-optical and microstructural properties of MgF_2-coatings deposited by IBS and PVD processes," Proc. SPIE **3738**, 436-445 (1999).
8. C. R. Clar, M. J. Dejneka, R. L. Maier, and J. Wang, "Extended lifetime excimer laser optics," US Patent 7242843 (2007).
9. R. J. Rafac, A. Lukashev, and W. Zhang Kevin, "Method and apparatus for stabilizing optical dielectric coatings," US Patent 20050008789 (2005).
10. V. Liberman, S. T. Palmacci, D. Hardy, M. Rothschild, and A. Grenville, "Controlled Contamination Studies in 193-nm Immersion Lithography," Proc. SPIE **5754**, 148–153 (2005).

11. V. Liberman, M. Switkes, M. Rothschild, S. T. Palmacci, J. H. C. Sedlacek, D. E. Hardy, and A. Grenville, "Long-Term 193-nm Laser Irradiation of Thin-Film-Coated CaF_2 in the Presence of H_2O," Proc SPIE **5754**, 646-654 (2005).
12. A. Duparre, R. Thielsch, N. Kaiser, S. Jakobs, K. R. Mann, and E. Eva, "Surface finish and optical quality of CaF_2 for UV lithography applications," Proc SPIE **3334**, 1048-1054 (1998).
13. A. Gatto, J. Heber, N. Kaiser, D. Ristau, S. Gunster, J. Kohlhaas, M. Marsi, M. Trovo, and R. P. Walker, "High-performance UV/VUV optics for the Storage Ring FEL at ELETTRA," Nucl. Instrum. Methods Phys. Res. A **483**, 357-362 (2002).
14. J. Ullmann, M. Mertin, H. Lauth, H. Bernitzki, K. Mann, R., D. Ristau, W. Arens, R. Thielsch, and N. Kaiser, "Coated optics for DUV excimer laser application," Proc SPIE **3902**, 514-527 (2000).
15. L. Martinu and D. Poltras, "Plasma deposition of optical films and coatings: A review," J. Vac. Sci. Technol. A **18**, 2619-2645 (2000).
16. L. Pawlowski, "Thick laser coatings: A review," J. Therm. Spray Technol. **8**, 279-295 (1999).
17. C. Amra, J. H. Apfel, and E. Pelletier, "Role of Interface Correlation in Light-Scattering by a Multilayer," Appl. Opt. **31**, 3134-3151 (1992).
18. V. Liberman, M. Rothschild, J. H. C. Sedlacek, R. S. Uttaro, and A. Grenville, "Excimer-laser-induced densification of fused silica: laser-fluence and material-grade effects on the scaling law," J. Non-Cryst. Solids **244**, 159-171 (1999).
19. E. Welsch, K. Ettrich, H. Blaschke, P. Thomsen-Schmidt, D. Schafer, and N. Kaiser, "Investigation of the absorption induced damage in ultraviolet dielectric thin films," Opt. Eng. **36**, 504-514 (1997).
20. H. Blaschke, W. Arens, D. Ristau, S. Martin, B. Li, and E. Welsch, "Thickness dependence of damage thresholds for 193-nm dielectric mirrors by predamage sensitive photothermal technique," Proc. SPIE **3902**, 242-249 (2000).
21. K. Ettrich, H. Blaschke, E. Welsch, P. Thomsen-Schmidt, and D. Schaefer, "UV-laser investigation of dielectric thin films," Proc. SPIE **2714**, 426-439 (1996).
22. H. Johansen and G. Kastner, "Surface quality and laser-damage behaviour of chemo-mechanically polished CaF_2 single crystals characterized by scanning electron microscopy," J. Mater. Sci. **33**, 3839-3848 (1998).
23. M. L. Protopapa, F. De Tomasi, M. R. Perrone, A. Piegari, E. Masetti, D. Ristau, E. Quesnel, and A. Duparre, "Laser damage studies on MgF_2 thin films," J. Vac. Sci. Technol. A **19**, 681-688 (2001).
24. J. Ferre-Borrull, A. Duparre, and E. Quesnel, "Roughness and light scattering of ion-beam-sputtered fluoride coatings for 193 nm," Appl. Opt. **39**, 5854-5864 (2000).
25. L. Escoubas, A. Gatto, G. Albrand, P. Roche, and M. Commandre, "Solarization of glass substrates during thin-film deposition," Appl. Opt. **37**, 1883-1889 (1998).
26. N. I. o. A. I. S. a. Technology, "Raman Spectra Database of Minerals and Inorganic Materials (RASMIN)" (National Institute of Advanced Industrial Science and Technology), retrieved http://riodb.ibase.aist.go.jp/rasmin/.
27. Q. Williams, *Infrared, Raman and Optical Spectroscopy of Earth Materials*, Mineral Physics and Crystallography - A Handbook of Physical Constants (American Geophysical Union Washington, D.C, 1995), Vol. AGU Reference Shelf 2, pp. 291-302.
28. N. Beermann, H. Blaschke, H. Ehlers, D. Ristau, D. Wulff-Molder, S. Jukresch, A. Matern, C. F. Strowitzki, A. Görtler, M. B. Gäbler, and N. Kaiser, "Long Term Tests of Resonator Optics in ArF Excimer Lasers," in *XVII International Symposium on Gas Flow and Chemical Lasers & High Power Lasers 2008*, 2008),

1. Introduction

The reliability and durability of optical instruments is limited by alteration or damage of the optical components and optics due to laser irradiation [1-7]. Thus, the influence of UV light on optical components is of great relevance for industrial applications such as lithography, metrology and medicine [8, 9]. In particular, deep ultra-violet (DUV) lasers damage calcium fluoride (CaF_2) windows after relatively short exposure times [2]. The damage appears as a white film on the exterior surface in the area of the beam profile. The composition and origin of this white material is so far unclear. A better understanding of the damage processes is needed to guide the development of improved UV optics with increased lifetime.

During the last few years, applications of deep ultra-violet (DUV) and vacuum ultra-violet (VUV) radiation have widely expanded. The current modes of laser operation in this optical range conform to the lithography sector, where the optical components are operating under well defined and closed working conditions at low energy densities [10, 11]. In contrast, the operating conditions in medical applications, photochemistry, micro-material machining, metrology, or life sciences are less well defined [5, 7, 12-14]. In these 'nonlithography' applications, the optics must sustain higher energy densities in poorly defined environments. Such demanding applications imply mechanical, thermal and chemical stress on the optical material, which result in lower product life times. In the worst case, the maintenance costs of optical components amount to 40 % of the total costs within tube durability. Thus, lifetime considerations for the optics must include endurance data of optic materials and thin films in ambient conditions.

Several experiments have been carried out to address this topic. Recent progress in the fabrication of plasma-assisted optical thin films and thick coatings for laser applications and their characteristics are summarized in [15, 16]. The importance of light scattering on multilayer coatings is highlighted in [17]. The dose-rate alteration (densification) of fused silica caused by an Excimer laser (ArF, 193 nm) was investigated to understand the effects of light fluence, pulse count and material grade [18]. The investigation of damage threshold for UV light and the behavior of dielectric thin films show dependence on substrate, thickness and layer material [19-21]. Light scattering on multilayer coatings is also very important [17]. The choice of polishing grades affects the quality of coated components. A better polishing grade increases the UV-laser (193 nm and KrF, 248 nm) damage threshold of CaF_2 [12, 22]. Thin magnesium fluoride (MgF_2) films deposited on CaF_2 [23] by electron-beam evaporation provide a very high UV-laser damage threshold of 9 J/cm^{-1} (focused 248 nm laser). Uncoated CaF_2 outcoupling optics showed damage of the external part of the laser resonator when exposed to energy densities of more than 50 mJ/cm^2 on the transmitting area. A similar damage of the high reflective mirror, however, never has been reported. A degradation of fused silica and CaF_2 with fluxes in the range from 0.2 to 4 mJ/cm^2 was reported in [2]. The chemical details of the damaging process remain unclear. In the following we show that the alteration of CaF_2 into $CaCO_3$ plays a major role in the damage mechanism.

2. Materials and methods

To determine the lifetime of an outcoupling mirror, we have performed marathon laser-irradiation tests for different optic substrates and thin film coatings. The samples were exposed to UV light from a water-cooled argon fluoride excimer laser (ArF, 193 nm, photon energy 6.42 eV, Coherent, Munich, Germany) with a solid-state pulse power switch, which has been tailored to purpose. The maximum laser pulse energy of 90 mJ was reached with the upper limit of the charging voltage of 1500 V with a maximal possible pulse frequency of 2 kHz and pulse duration of about 12 ns (FWHM). With a beam profile of 10 × 3.5 mm^2, an energy density of more than 240 mJ/cm^2 could be achieved. All measurements were carried out in an energy-stabilized mode at 1 kHz in continuous operation, where the high voltage was regulated to an output pulse energy of 28 mJ (80 mJ/cm^2). The pulse energy was measured externally by a power meter (PM 150-50, Coherent, USA). A stream of filtered laser gas minimized the deposition of dust that was generated from electrode abrasion due to the high voltage discharge on the resonator internal side.

The outcoupling test optics (diameter 38.1 mm, 5 mm thickness) were loaded into the mirror mount for each experiment and sealed against the atmosphere by an O-ring. The optical mount was designed to allow for the access to the sample for direct temperature measurements on the mirror at distances of 1 mm, 3 mm, and 6 mm from the laser beam. The generation of oxygen radicals and ozone in front of the optic was minimized by flushing the external optical path with high-purity nitrogen. All components were made from stainless steel. Contamination was prevented by carefully handling components and optics in a clean environment. The dielectric reflection coatings (>98 % for the high reflective mirror and 25 % for the outcoupling mirror) were on the resonator's internal side with contact to the fluoric laser gas. One vendor provided all the polished CaF_2 substrates. The optical coatings on the substrates were fabricated by different manufacturers. Due to the variety of coating producers, the coatings differed in layer design and production.

Various state-of-the-art optics were tested until 250 million laser pulses were reached or damage occurred. In a second set of experiments, the measurements were repeated with improved optics, which were produced with improved manufacturing parameters for thin film deposition, layer design, and crystal polishing. These mirrors were stepwise tested until damage occurred. After each step, the outcoupler was inspected optically and the test was continued if there was no sign of alteration otherwise the damage was investigated.

We used a confocal Raman microscope alpha300 R (WITec, Ulm, Germany) for Raman measurements. The system used a standard 100× microscope objective (Air, NA = 0.90, working distance 0.26 mm, Nikon, Düsseldorf, Germany) for diffraction limited focusing and collecting the scattered light. A frequency doubled Nd:YAG (532 nm) laser was used for excitation. The scattered light was analysed with a lens based spectrometer, equipped with an 1800 lines/mm grating.

Energy dispersive X-ray spectroscopy (EDX), integrated in a transmission electron microscope (TEM/STEM JEM-3010, JEOL) was used to characterize elements on top and below the damaged mirror surface to clarify the chemical modification of the bulk material. The EDX measurements were supported by time-of-flight secondary ion mass spectrometry (ToF-SIMS; IONTOF TOF.SIMS5-300). Transmission Electron Microscopy (TEM CM20, Phillips) was used to characterize the structure of subsurface damaged regions. A focused ion beam (FIB, NVision, ZEISS) technique was used for TEM sample preparation and subsurface 3D volume reconstruction.

3. Results and discussion

3.1 Damage formation

Various UV transparent substrates were tested prior to the experiments: barium fluoride (BaF_2), UV grade fused silica, lithium fluoride (LiF), samarium fluoride (SmF_3), MgF_2 and CaF_2. All crystals showed color centers after a short exposure time, with the only exception of CaF_2. Thus, we focused on CaF_2 as a substrate material because it showed the lowest tendency of color center formation [15, 24, 25].

All tests showed similar temperature characteristics. Directly after starting the irradiation test, the temperature increased and stabilized within a few hours at about 70 °C (1 mm from laser beam boundary), 65 °C (3 mm) and 55 °C (6 mm). The laser tube temperature was in the range of 40 °C to 45 °C, depending on the discharge voltage for laser power control. The window temperature inside the beam profile could not be measured. It was estimated to be much higher than 150 °C. Shortly before damage occurred, the temperature on the exterior surface rapidly increased to 150 °C (1 mm) within some million pulses (ca. 1h).

For all samples, initial external surface damage occurred at the outer boundary of the transmitted laser beam profile of 3.5×10 mm^2. Once damage occurred, the damaged area grew rapidly towards the center of the beam profile. The morphology and size of the damage was similar on all samples. There was only a slight variation in number of laser pulses until damage occurred. In the following, we will discuss the chemical alteration process of two typical samples: one uncoated and one with a protective MgF_2 coating.

3.2 Raman analysis

The damaged area of a mirror without protection layer was characterized with the confocal Raman microscope. Figure 1(a) shows an optical micrograph of the damage area, whose surface size was 3.5×10 mm^2. The damage appears as a white contamination on the surface. Figure 1(b) shows a detail of the damage boundary.

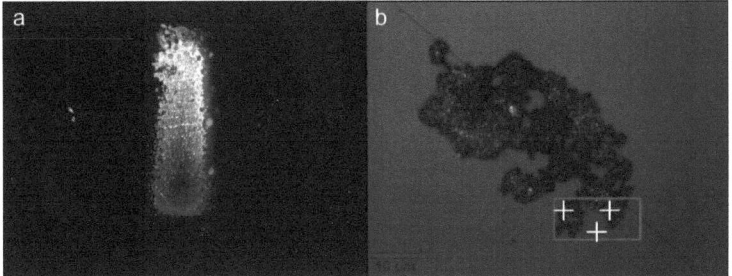

Fig. 1. Surface damage of an uncoated CaF_2 mirror. (a) Optical image of the entire damaged area. Size of the beam profile (3.5×10 mm^2). (b) Detailed optical image of a structure at the boundary of the damage area. The white crosses give the position of the Raman spectra in Fig. 2. The red rectangle is the area of interest of the micro-Raman image in Fig. 3.

Spectra measured with a 10 s integration time on the damaged and undamaged material are shown in Fig. 2. Within the spectra, five intense sharp peaks are visible together with fluorescence in a range of 150-500 cm^{-1}. One peak at 330 cm^{-1} was present in all spectra, the others (155 cm^{-1}, 284 cm^{-1}, 713 cm^{-1} and 1089 cm^{-1}) were only found in the damaged areas.

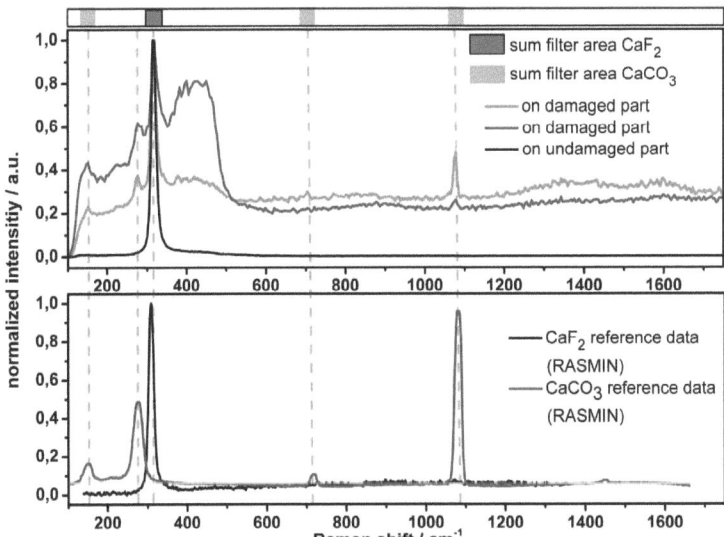

Fig. 2. (Top) Single Raman spectra measured with 10 s integration time as indicated in Fig. 1. (Bottom) Reference Raman spectra of pure crystals from the RASMIN database [26]. The characteristic peaks for CaF_2 (330 cm^{-1}) and $CaCO_3$ (155 cm^{-1}, 284 cm^{-1}, 713 cm^{-1} and 1089 cm^{-1}) are clearly visible in the measured spectra. The sum filter window used for CaF_2 (bright blue) and $CaCO_3$ (grey) mapping are indicated.

These peaks correspond to CaF_2 (330 cm^{-1}) and calcite (crystalline $CaCO_3$) [26, 27]. Table 1 lists the peak assignment and the corresponding mode symmetry and origin. There is a very good agreement of the measured peaks with the reference data. In the damaged areas, some fluorescence with minor peaks occurred in the spectra, which is typical of organic contaminants.

Table 1. Peak assignment for Figs. 2 and 3[a].

Raman shift (cm^{-1})	Mode symmetry	Material
155	lattice mode	CaCO$_3$ (calcite)
284	lattice mode	CaCO$_3$ (calcite)
713	In-plane bend	CaCO$_3$ (calcite)
1089	CO$_3$ symmetric stretch	CaCO$_3$ (calcite)
1435	CO$_3$ asymmetric stretch	CaCO$_3$ (calcite) not visible
330	lattice mode	CaF$_2$ (calcium fluoride)

[a] according to RASMIN Database and Mineral Physics and Crystallography: A Handbook of Physical Constants, Vol. 2, pp. 291-302.

The images in Figs. 3(a) and 3(b) show the lateral distribution of calcium fluoride and crystalline calcium carbonate (calcite). The images consist of 160 spectra per line and 80 lines per image with a size of 18.6 × 8.6 µm². Each spectrum was integrated for 1 s. The spectral images were filtered by summing up the counts in a selected interval around a Raman peak as indicated in Fig. 2 (top).

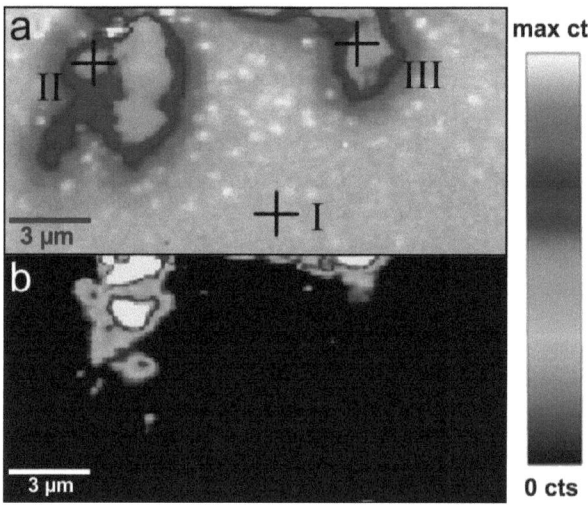

Fig. 3. Micro-Raman analysis of the exterior surface as indicated in Fig. 1 (image size: 18.6 × 8.6 µm²) .The images consist of 160 spectra per line and 80 lines per image. (a) Map of the CaF$_2$ lattice mode (filter window: 315 – 355 cm^{-1}) and the (b) CO$_3$ symmetric stretch mode (1080 – 1110 cm^{-1}). With decreasing CaF$_2$ peak intensity the peak intensity of CaCO$_3$ increases. The size of the CaCO$_3$ (calcite) particles is 2 – 3 µm.

The lateral intensity distribution of CaF$_2$ lattice mode at 330 cm^{-1} (filter window: 315 – 355 cm^{-1}) shows a higher intensity in the undamaged area and a decreased intensity in the damaged part. Simultaneously, the intensity of the CO$_3$ symmetric stretch peak (1089 cm^{-1}, filter window 1080 – 1110 cm^{-1}), is higher in the damaged area. The size of the CaCO$_3$ (calcite) particles could be resolved to a few micrometers. The calcite crystals were surrounded by organic material (fluorescence at the lower wave number ranges) appearing as dark blue in Fig. 3(a) and bright blue in Fig. 3(b).

Vertical slices were measured for subsurface investigation. Figure 4(a) shows the optical image of a damaged area. The black line indicates the deep scan location were a slice was acquired perpendicular to the surface. The total scanning height was 8 µm starting at 3 µm above the surface with a length of 15 µm. 40 lines with 150 points were acquired; each spectrum was integrated for 1.5 s. Figure 4(b) shows the sum filtered CaF_2 lattice mode image (330 cm^{-1}) and Fig. 4(c) a map of the CO_3 symmetric stretch mode (1089 cm^{-1}). The image indicates that $CaCO_3$ occurs in a depth 1 µm beneath the surface (considering the different indexes of refraction). Additionally, the $CaCO_3$ has grown to a height about 0.5 µm above the surface.

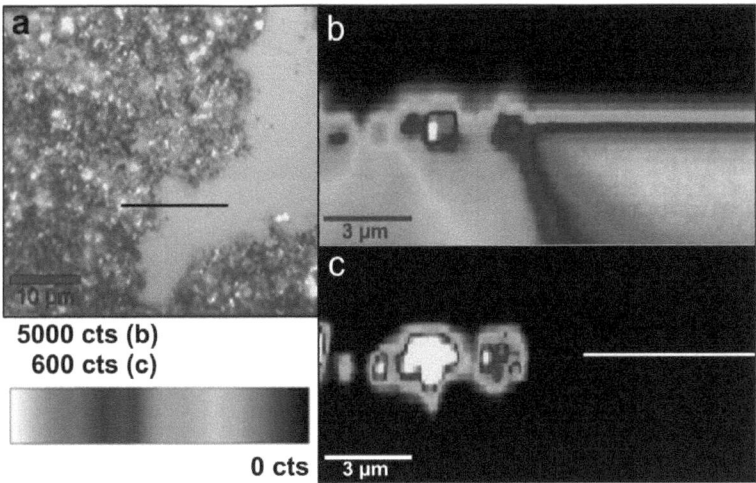

Fig. 4. Surface damage of an uncoated CaF_2 mirror. (a) Optical image of the damage area. The black line indicates the location of the vertical profile; Raman images of the (b) sum filtered CaF_2 lattice mode (315 – 355 cm^{-1}) and the (c) sum filtered CO_3 symmetric stretch mode (1080 –1110 cm^{-1}). The white line in (c) indicates the sample surface. The altered material was found at about 0.8 µm above the CaF_2 surface and up to 2 µm below. (b) and (c): Raman images, 1.5 s integration time, image size: 15 µm × 8 µm^2, starting 3 µm above the surface.

Figure 5 shows a vertical profile of a mirror with an exterior protection layer of MgF_2 (<100 nm). The optical image in Fig. 5(a) shows a part of the damaged area, which is covered with debris from the protection layer. The black line indicates the position of a vertical profile. The 25 µm long slice was taken from 3 µm above to 5 µm below the surface. The images consist of 200 points per line and 30 lines per image with an integration time of 1.5 s for each spectrum. A Raman map of the CaF_2 lattice mode is shown in Fig. 5(b). Figure 5(c) shows a map of the MgF_2 lattice mode; Fig. 5(d) shows the CO_3 symmetric stretch and some artifacts. The white arrows indicate calcite crystals. It is evident, that the CaF_2 and MgF_2 signals are decreased on the altered part. This implies that the protective coating has been removed and that CaF_2 was altered to $CaCO_3$.

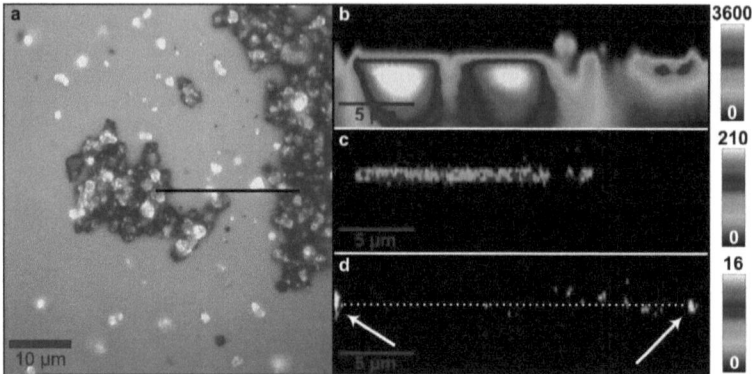

Fig. 5. Surface damage on a CaF_2 mirror with MgF_2 protection layer. (a) Optical image of the damage area. The black line indicates the location of the vertical profile. Raman maps of the (b) CaF_2 lattice mode, the (c) MgF_2 lattice mode, and (d) the CO_3 symmetric stretch mode. The white arrows indicate calcite. The spectra were acquired with 1.5 s integration time. The slice was taken from 3 µm above to 5 µm beneath the surface.

3.3 SEM /TEM/EDX, and ToF-SIMS analysis

Scanning electron microscope (SEM) images (Fig. 6(a)) of an exterior unprotected mirror show dendric pyramidal crystals grown on the surface. A surface damage with a height of about 1 µm and a depth of about 1 to 2 µm was observed (SEM cross-section, Fig. 6(b)). The crystals nucleated independently and grew in a 3-fold symmetric pyramids. All the trees have the same orientation of the shape, which should be defined by crystallographic orientation of the matrix. Electron diffraction and EDX data show that these pyramids are composed of CaF_2. Cross-sectional TEM imaging (Fig. 6(c)) reveals cracks propagating 1-2 µm beneath the surface. The cracks occur although the matrix beneath is perfectly crystalline, as indicated by bending contours (dark curved lines). 100-500 nm sized crystals were observed on dark field images (not shown here), in SEM cross-sectional views as slightly darker areas around the cracks and on EDX maps (Fig. 6(c) insert) as inclusions in the matrix at the edges of cracks (marked as V at Fig. 6(c)). These inclusions either represent slightly rotated grains (as evidenced by electron diffraction) or an oxygen rich phase (as evidenced by EDX).

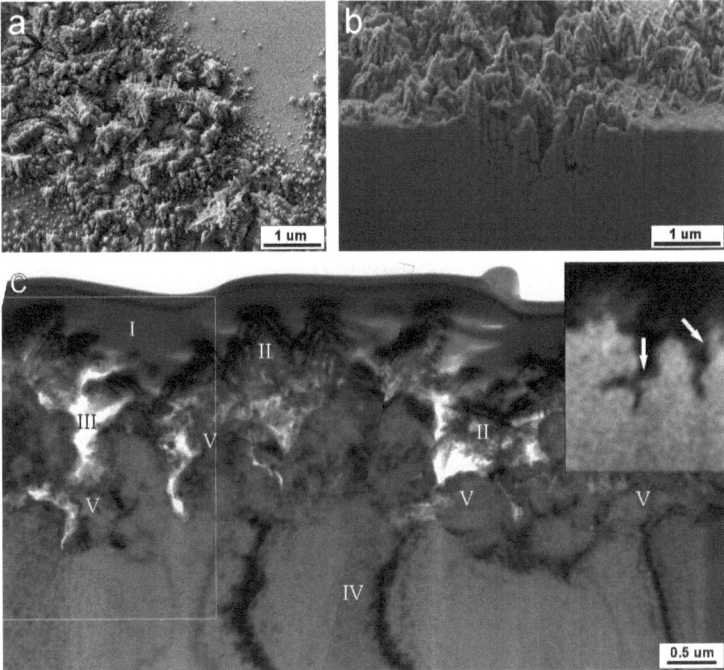

Fig. 6. Surface damage on an unprotected CaF$_2$ mirror. SEM (a, b), TEM (c) images and EDX composition map (c insert). (a) SEM image of the surface at the damaged region. (b) Cross-sectional SEM image and movie (Media 1) at the edge of the damaged region. Cutting was done by FIB technique at NVision instrument (ZEISS, Oberkochen). (c) TEM collage image of the subsurface region: (I) deposited protecting Pt/C layer, (II) dendrites with protective Au deposited layer (dark), (III) pores (cracks), (IV) bulk CaF$_2$ material, dark curved lines on the matrix indicate perfect crystallinity, (V) inclusions of the material distinct from the matrix (these regions are seen in dark field TEM images (not shown here) and at EDX maps). Inset in (c) is a RGB (Ca, F, O) composite image of the EDX elemental maps of the region marked by a rectangle. The oxygen peak is overlain by a minor Au peak, resulting in an artefact surface layer at oxygen map. The maximum oxygen content was found at the positions indicated by the arrows. Fluorine and gold are absent in these areas.

Using a sequence of images (Fig. 6(b)) a tomographic reconstruction of subsurface volume was made (Fig. 7). The reconstruction revealed a peculiar porous structure with pores having a shape of conical structures or hemi-spheres and the cracks developing from the tops of the conical structures.

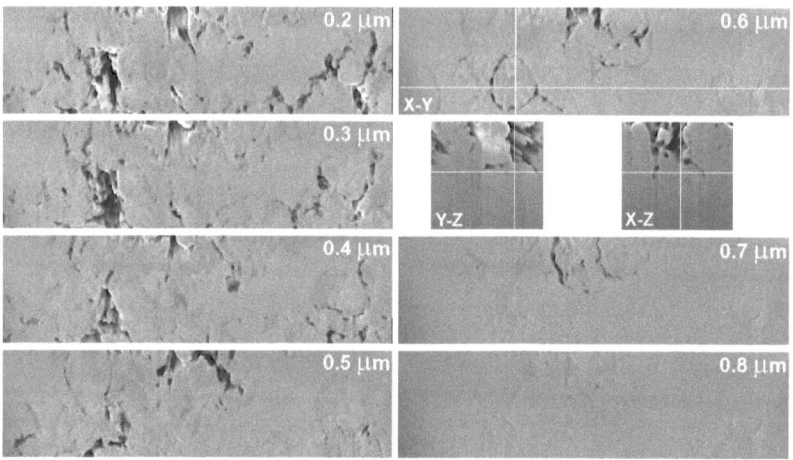

Fig. 7. Tomographic reconstruction of the subsurface region. Images represent subsequent slices parallel to the surface going into the depth of the sample; the depth is indicated on corresponding images. Below 0.6 µm slice cuts in two other complementary directions are shown, revealing a 3D conical shape of the round pore and a crack on top of it.

ToF-SIMS analysis showed aliphatic and aromatic hydrocarbons, fatty acids, and nitrogen and oxygen-bearing hydrocarbons on the entire sample surface. In the inner part, nitrogen and oxygen-bearing hydrocarbons (in particular C_3H_8N, $C_{18}H_{40}N$, $C_{20}H_{44}N$) were found together with triglycerides. Finger prints can be barred as the origin for the hydrocarbons due to the usage of clean room gloves. A deep profile analysis of the damage area (300×300 µm^2) showed higher intensities of Ca_2OF, Ca_xO_y and $CaOH$. An elevated $CaCO_3$ intensity could not be observed. This was probably due to the large scanning area and small amount of $CaCO_3$.

3.4 Discussion

The damaged area of high flux, illuminated DUV outcoupling optics was investigated with different analytic methods. The damage started at the outer boundary of the transmitted laser beam and grew towards the center of the beam profile. The exterior side was roughened and small amounts of CaF_2 were altered into calcite (crystalline $CaCO_3$). Additionally, calcium oxide was detected, which was likely an intermediate. As a consequence, the transmitted laser power was decreased due to scattering losses and adsorption respectively.

A possible deterioration process is discussed in detail in [28]. Briefly, the alteration of the optics can be attributed to the influence of the high-energy photons flux on defects located at the interface between substrate and coating.

The exchange reaction is supposed to follow:

$$CaF_2 + 2H_2O + CO_2 \Rightarrow Ca(OH)F + HF + CO_2 \Rightarrow$$
$$\Rightarrow Ca(OH)_2 + 2HF + CO_2 \Rightarrow CaCO_3 + 2HF + H_2O \quad (1)$$

This assumption is supported by the observation that the temperature on the boundary and in the center of the beam profile is very high (>150 °C) and the UV photon energy (6.42 eV). The very small amount of emerging HF is rapidly removed by the external flushing with nitrogen. Calcium oxide hydrates $Ca(OH)_2$ were probably generated in an exothermic reaction due to presence of water. Minor amounts of water are present on surfaces in ambient conditions. Such a reaction could be partially responsible for the increase in temperature, which was observed shortly before the damage occurred. Together with increased light absorption, such a reaction expedites the damaging process.

4. Conclusion

The laser-induced damage on the exterior side of outcoupling CaF_2 mirrors was investigated. The mirror was subject to a transmitted ArF UV-laser beam with an energy density of 80 mJ/cm^2. The irradiation caused the surface to roughen and to partially alter CaF_2 into $CaCO_3$. The exchange was driven by temperature and the photo-induced chemistry due to the UV laser light. The alteration started at the edge of the beam profile where the highest temperature gradients occur. Crystalline $CaCO_3$ (calcite) within the damaged area could be observed by confocal Raman spectroscopy. Further surface analysis with TEM/EDX confirmed this result. ToF-SIMS also revealed calcium oxide, which is very likely a precursor for the alteration to $CaCO_3$. Calcium oxide cannot be detected by Raman spectroscopy and thus went undetected. The $CaCO_3$ crystals were embedded in a variety of organic materials, which were identified using surface analytical techniques. High energetic photons flux induces photochemical reactions of imperfections located at the interface between substrate and coating or with contaminants from the environment.

The results indicate that a better protection of the surface is required in order to avoid the alteration of CaF_2. Such a protection could be realized by additional buffer layers, improved optical materials, advanced deposition processes and for coating designs.

Acknowledgment

We gratefully acknowledge financial support by the German Federal Ministry of Education and Research (BMBF) for founding the cooperative project FLUX (FKZ: 13N8937). We thank VDI-TZ, division "Optische Technologie", for supporting the subproject "Untersuchungen der Dauerbetriebs-festigkeit von beschichteten optischen Bauteilen im Resonator von Excimerlasern bei 193 nm". We are grateful to Prof. Rettenmeyer (University Jena) for providing TEM/EDX - and to Dr. Horn (Zeiss, NTS) for FIB access. Furthermore we thank Guido Winkler and Elke Tallarek, tascon GmbH for ToF-SIMS analyses.

7.5 Onset of the optical damage in CaF_2 optics caused by deep-UV lasers

Michael Bauer, Martin Bischoff, Thomas Hülsenbusch, Ansgar Matern, and Robert W. Stark, Norbert Kaiser

Optics Letters Vol. 34, Iss. 24, pp. 3815-3817

Permanent weblink:
http://dx.doi.org/10.1364/OL.34.003815

Onset of the optical damage in CaF_2 optics caused by deep-UV lasers

Michael Bauer,[1,]* Martin Bischoff,[3] Thomas Hülsenbusch,[2] Ansgar Matern,[2] Robert W. Stark,[1] and Norbert Kaiser[3]

[1]*Center for NanoScience (CeNS) and Department of Earth and Environmental Sciences, Ludwig-Maximilians-Universität München, Theresienstrasse 41, 80333 Munich, Germany*
[2]*Coherent GmbH, Zielstattstraße 32, 81379 Munich, Germany*
[3]*Fraunhofer-Institut für Angewandte Optik und Feinmechanik IOF, Albert-Einstein-Strasse 7, 07745 Jena, Germany*
Corresponding author: michael.bauer@lrz.uni-muenchen.de

Received August 31, 2009; accepted October 30, 2009;
posted November 13, 2009 (Doc. ID 116524); published December 7, 2009

The exterior sides of calcium fluoride (CaF_2) outcoupling mirrors are damaged by ArF laser light irradiation with high pulse-energy densities (80 mJ/cm^2). The damage is generated by a partial alteration of the CaF_2 substrate to calcite. The CaF_2 decomposition is driven by photochemical processes due to the UV light and the presence of water vapor and is supported by elevated temperatures within the laser beam transmitting area. Small filaments act as starting points for the decomposition process, where kerogenous carbon and calcite can occur. © 2009 Optical Society of America

OCIS codes: 140.3330, 230.4040, 140.2180, 260.7200.

Intense ultraviolet (UV) laser light used in various areas such as lithography, metrology, medicine, and high-power industrial applications can degrade optical components [1–4]. In particular, high-flux deep-UV (DUV) lasers generate damage structures on calcium fluoride (CaF_2) windows already after rather short exposure times [3]. Within the surface area of the transmitted laser beam a white film forms on the exterior surface [5]. The high-energy photon flux induces photochemical reactions at imperfections or of contaminants from the environment with the window material [6]. Controlled studies on the influence of contaminants and the presence of water show that uncoated CaF_2 is attacked by water within several hours [2,7]. Thus, a precise knowledge of the early stages of material alteration is required to develop optics and protective coatings with increased lifetime.

Recent investigations showed that a partial decomposition of the CaF_2 bulk material into crystalline calcium carbonate ($CaCO_3$, calcite) is responsible for the damage [6,8]. The decomposition is mainly driven by photochemical processes in the presence of water vapor. The main parameters are high-UV photon energy (6.42 eV, 193 nm), short pulse duration (~ 12 ns), high pulse power density, and elevated temperatures. The temperature of the mirror rises shortly before the material alteration becomes visually evident. An increase from about 70 °C to more than 150 °C was measured at 1 mm distance from the irradiated surface area. Chemical surface analysis revealed oxygen-rich phases within the damage area including calcium oxide (CaO), which is very likely a precursor for the alteration to $CaCO_3$. Both CaO and $CaCO_3$ adsorb in the UV region and thus may expedite the damage process. In addition, a variety of organic materials could be observed such as nitrogen- and oxygen-bearing hydrocarbons and triglyceride. The first stages of the degradation process, however, are not yet understood.

In the following we show that the damage process is initiated by small dendritic filaments growing on the exterior surface within the surface area of the transmitted laser beam. To investigate this process, CaF_2 outcoupling test mirrors (diameter 38.1 mm, thickness 5 mm, no protective coating) were loaded into the mirror mount and sealed with an O-ring. The deposition of dust produced by electrode abrasion during discharge was minimized with a stream of filtered laser gas on the resonator's internal side. To prevent the generation of oxygen radicals and ozone in front of the optics, the external optical path was flushed with high-purity nitrogen. All components were made from stainless steel. Contaminations were avoided by carefully handling the components in a clean environment. The samples were exposed to UV light from an argon fluoride excimer laser [ArF, 193 nm, 6.42 eV, 12 ns pulse duration (FWHM), Coherent, Munich, Germany]. All measurements were carried out in an energy-stabilized mode at 1 kHz in continuous operation at an output pulse energy of 28 mJ (80 mJ/cm^2, >6 MW/cm^2 pulse peak power).

The laser emission was stopped after approximately 100 million laser pulses, and the outcoupling mirror was examined. A visual inspection revealed damage structures smaller than those found after an exposure to more than 250 million laser pulses [8]. Small individual filament structures as well as filaments that were connected to damage structures occurred after 100 million pulses. To clarify the damage mechanism we inspected these structures in more detail.

For compositional mapping the surface was imaged with a confocal Raman microscope, alpha300 R (WITec, Ulm, Germany). The system used a standard $100\times$ microscope objective (Air, NA=0.90, working distance 0.26 mm, Nikon, Düsseldorf, Germany) for diffraction-limited focusing and collecting the scattered light from a frequency-doubled Nd:YAG (532 nm, 22 mW) laser. The inelastically scattered

0146-9592/09/243815-3/$15.00 © 2009 Optical Society of America

light was analyzed with a lens-based spectrometer equipped with an 1800 lines/mm grating. At each image point, a full spectrum was recorded with an integration time of 1 s if not stated otherwise. Compositional maps were calculated by integrating the counts over a defined spectral range.

An optical micrograph of a more advanced damage with protruding filaments is shown in Fig. 1. Two of the filaments were crossing each other, with a small particle at the intersection point. Confocal Raman maps were acquired in this region of interest (not shown). The white arrows mark the positions where typical spectra with a signature of CaF_2, calcite, or kerogenous carbon were selected for presentation. Figure 2 shows spectra of (a) CaF_2, calcite, and kerogenous carbon; (b) CaF_2 and calcite only; and (c) CaF_2 and organics. The spectra were obtained each with 10 s integration time. In more detail, the spectrum (a) consists of the Raman bands of CaF_2 (lattice mode at 330 cm^{-1}) and calcite (lattice mode at 155 cm^{-1} and 284 cm^{-1}, in-plane bend at 713 cm^{-1}, and CO_3 symmetric stretch mode at 1089 cm^{-1}). The weaker CO_3 asymmetric stretch mode at 1435 cm^{-1} is masked. In addition to these assigned peaks, Raman peaks at around 1350 cm^{-1}, 1600 cm^{-1}, 2700 cm^{-1}, and 3000 cm^{-1} were present. These peaks can be attributed to the kerogenous carbon bands D1, G, S1, and S2. Kerogenous carbon is often found in organic remains in geological sediments [9,10]. The local temperature during kerogen formation can be deduced from the peak shape and height of the D and G (R2 ratio) [9]. Here, the analysis indicates a temperature of about 300°C to 350°C within the beam transmitting area, which is compatible with the measured temperature increase at 1 mm distance. In the spectrum taken on the particle [Fig. 1(b)], the D1 and S1 bands of kerogen are absent and the individual peaks are weaker than in (a). The particle material can be identified as calcite. Spectrum (c) is dominated by fluorescence from organic matter, schematically indicated by the lower curve (red online). Two additional peaks were present at 330 cm^{-1} (CaF_2 lattice

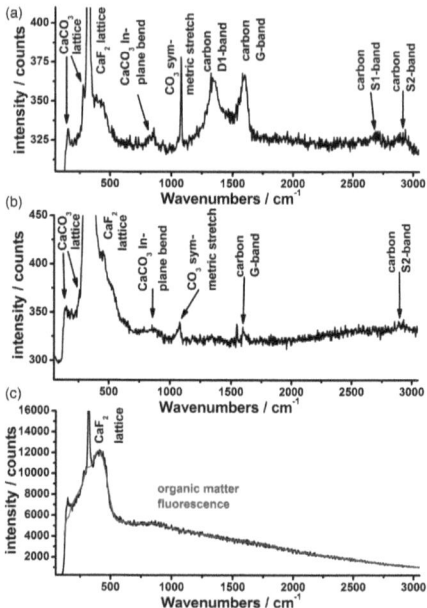

Fig. 2. (Color online) Raman spectra on the positions marked in Fig. 1. They were obtained with an integration time of 10 s per spectrum. (a) $CaCO_3$ (calcite), CaF_2, and kerogenous carbon Raman peaks (D1, G, S1, S2 band). (b) Spectrum obtained on the calcite particle located on the filament. (c) Spectrum dominated by fluorescence. Fluorescence of organic matter is indicated by the lower curve (red online).

mode) and 147 cm^{-1}. The absence of an intense CO_3 symmetric stretch suggests that the peak at 147 cm^{-1} cannot be assigned to the CO_3 lattice mode (155 cm^{-1}). This fact in turn implies that calcite was absent at this location.

The optical micrograph in Fig. 3(a) shows another typical filament with small particles within the structure. Figure 3(b) was generated by integration of the most intense calcite peak (CO_3 symmetric stretch) in the range from 1040 cm^{-1} to 1110 cm^{-1}. The image is an overlay of the CO_3 distribution in dark gray (II, red online) with the distribution of organics in light gray (I, green online). It is obvious that the filaments consisted of organic material with some calcite constituents. Figure 3(c) shows an image calculated by a sum filter for the range from 1280 cm^{-1} to 1420 cm^{-1} corresponding to the D1 band of kerogenous carbon (yellow online). In the map, the kerogenous material is located near the position where the signal of the calcite CO_3 symmetric stretch peak prevails. The growth orientation follows the filament structure as evident in the optical image. The vertical slice in Fig. 3(d) was obtained at the positions marked with the white line in Fig. 3(a). The signal from the CaF_2 sub-

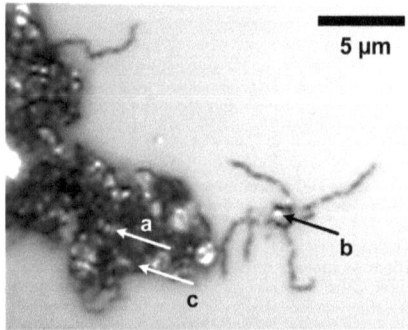

Fig. 1. Optical micrograph of an advanced damage with filaments and calcite particles on top. The arrows mark the position of the spectra in Fig. 2.

ment overgrown by the damage in Fig. 1 suggest a damage growth that started on the filaments and then followed the structure.

In summary, our data indicate that the decomposition of CaF_2 substrate starts at filamentlike structures on the exterior surface that grow to a distinctive damage. Raman scattering at kerogenous carbon, likely formed by altered organic contaminations, indicates a maximum local mirror temperature of more than 300°C within the laser beam transmitting surface area during irradiation. The alteration of CaF_2 seems to progress as a dendritic growth sprouting on the filaments and propagating along them. Calcite particles emerge at the filaments, which act as starting points for the subsequent degradation process. This process is supported by an increased light absorption and increased scattering caused by the roughened surface.

We gratefully acknowledge financial support by the German Federal Ministry of Education and Research (BMBF) for founding the cooperative project FLUX (FKZ: 13N8937).

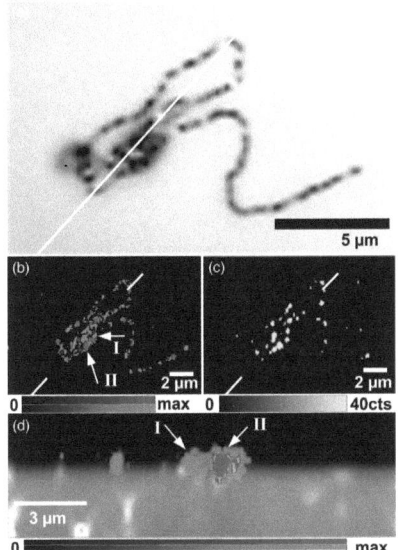

Fig. 3. (Color online) Compositional mapping of a filamentous damage structure. (a) Optical micrographs of an individual filament. The white line marks the position of the vertical slice in (d). (b) Intensity distribution of the CO_3 symmetric stretch peak (II, dark gray; red online) with organics (I, light gray; red online) and (c) carbon D1 band (yellow online). (d) Vertical section as indicated in (a). CaF_2 is shown in gray scale, the CO_3 symmetric stretch peak in dark gray (II, red online) fluorescence from organics in light gray (I, green online). Image size: (b), (c) 12 µm ×17.5 µm, 150 lines×100 points; (d) 5 µm×16 µm, 25 lines×140 points.

strate (integration from 315 cm^{-1} to 355 cm^{-1}) is displayed in gray scale, the signals assigned to the CO_3 symmetric stretch peak (1040 cm^{-1} to 1110 cm^{-1}) and organics in dark gray (II, red online) and light gray (I, green online), respectively. The particles were about 1 µm high and were located only at the surface. They were embedded in a jagged CaF_2 surface matrix [8]. The single filament in Fig. 3 and the fila-

References

1. N. L. Boling, M. D. Crisp, and G. Dube, Appl. Opt. **12**, 650 (1973).
2. V. Liberman, S. T. Palmacci, D. Hardy, M. Rothschild, and A. Grenville, Proc. SPIE **5754**, 148 (2005).
3. V. Liberman, M. Rothschild, J. H. C. Sedlacek, R. S. Uttaro, A. Grenville, A. K. Bates, and C. Van Peski, Opt. Lett. **24**, 58 (1999).
4. H. Johansen and G. Kastner, J. Materials Sci. **33**, 3839 (1998).
5. H. Blaschke, N. Beermann, H. Ehlers, D. Ristau, M. Bischoff, D. Gabler, N. Kaiser, A. Matern, and D. Wulff-Molder, Proc. SPIE **7132**, 7132A (2008).
6. N. Beermann, H. Blaschke, H. Ehlers, D. Ristau, D. Wulff-Molder, S. Jukresch, A. Matern, C. F. Strowitzki, A. Görtler, M. Bischoff, D. Gäbler, and N. Kaiser, Proc. SPIE **7131**, 713117 (2008).
7. V. Liberman, M. Switkes, M. Rothschild, S. T. Palmacci, J. H. C. Sedlacek, D. E. Hardy, and A. Grenville, Proc. SPIE **5754**, 646 (2005).
8. M. Bauer, M. Bischoff, S. Jukresch, T. Hülsenbusch, A. Matern, A. Görtler, R. W. Stark, A. Chuvilin, and U. Kaiser, Opt. Express **17**, 8253 (2009).
9. O. Beyssac, B. Goffe, C. Chopin, and J. N. Rouzaud, J. Metamorph. Geol. **20**, 859 (2002).
10. O. Beyssac, B. Goffe, J. P. Petitet, E. Froigneux, M. Moreau, and J. N. Rouzaud, Spectrochim. Acta **59**, 2267 (2003).

Appendix

Draft: Raman Spectroscopy of laser induced oxidation of titanomagnetites

Michael Bauer, Polina Davydovskaya, Marek Janko, Melanie Kaliwoda, Nikolai Petersen, Stuart Gilder, Robert W. Stark

Manuscript submitted

Raman spectroscopy of laser induced oxidation of titanomagnetites

Michael Bauer[1,2], Polina Davydovskaya[1,2], Marek Janko[1,2], Melanie Kaliwoda[3], Nikolai Petersen[1], Stuart Gilder[1], Robert W. Stark[2,4,5,*]

[1] Department of Earth and Environmental Sciences, Ludwig-Maximilians-Universität München, Theresienstr. 41, 80333 Munich, Germany
[2] Center for NanoScience (CeNS), Schellingstr. 4, 80799 Munich, Germany
[3] Mineralogical State Collection, LMU, Theresienstrasse 41, 80333 Munich, Germany
[4] FB Material- und Geowissenschaften, TU Darmstadt, Petersenstr. 32, 64287 Darmstadt, Germany
[5] Center for Smart Interfaces, TU Darmstadt, Petersenstr. 32, 64287 Darmstadt, Germany

*corresponding author: Phone: +49 6151 16 2109, Fax: +49 6151 16 72090, e-mail: stark@csi.tu-darmstadt.de

Abstract: Titanomagnetites are important carriers of magnetic remanence in nature and can track redox conditions in magma. The titanium concentration in magnetite bears heavily on its magnetic properties, such as saturation moment and Curie temperature. On land and in the deep ocean, however, these minerals are prone to alteration which can mask the primary magnetic signals they once recorded. Thus, it is essential to characterize the cation composition and oxidation state of titanomagnetites that record the paleomagnetic field. Raman spectroscopy provides a unique tool for both purposes. Nonetheless, the heat generated by the Raman laser can itself induce oxidation. We show that the laser power threshold to produce oxidation decreases with increasing titanium content. With confocal Raman spectroscopy and magnetic force microscopy on natural and synthetic titanomagnetites, non-destructive Raman imaging protocol was established. We applied this protocol to map out the composition and magnetization state within a single ex-solved titanomagnetite grain in a deep sea basalt.

Keywords: titanomagnetite, alterations, local heating, MFM, confocal Raman

1. Introduction

The composition of magnetic minerals provides information on the conditions of rock genesis and alteration.[1] Minerals from the titanomagnetite solid solution series ($Fe_{3-x}Ti_xO_4$, $0 < x < 1$) are of particular relevance because they carry the remanent magnetization of oceanic basalts.[2, 3] The composition of titanomagnetite is close to x = 0.6 with a Curie temperature of 140 °C when formed at a spreading ridge.[1] Their equilibrium oxygen fugacity is on the order of 10^{-7} atm, implying a strong tendency to oxidize under sea-floor conditions.[4] To characterize the composition and oxidation state of titanomagnetite Raman imaging techniques are valuable tools.[5] Raman spectroscopic characterization and compositional analysis of rock-forming minerals will also be an important tool in future planetary missions.[6, 7] Iron oxides, however, are poor Raman scatterers that strongly absorb in the range of the wavelengths of typical excitation lasers and high laser intensities are required to obtain an adequate signal.[1, 8] Thus, the heat generated by the strongly focussed Raman laser may lead to alteration. To obtain reliable Raman data on titanomagnetite, the measurement conditions need to be optimized to reduce the measurement time without altering of the specimen.

Stoichiometric titanomagnetites ($Fe_{3-x}Ti_xO_4$) can be synthesized in a wide range of compositions.[9] Magnetite, i.e. x = 0, also referred to as TM00, is a special case of ferrite with the general formula $Fe^{2+}Fe_2^{3+}O_4$ corresponding to $FeO \cdot Fe_2O_3$ (simplified Fe_3O_4) with an inverse spinel structure.[10] TM60 (x = 0.6) represents an idealized natural titanomagnetite with the composition $Fe_{2.4}Ti_{0.6}O_4$. The long range order in synthetic TM60 single crystals is a function of nonstoichiometry with higher cation vacancy concentrations producing a more random cation distribution.[11] The magnetic stability of synthetic titanomagnetite can be increased by oxidation and subsequent annealing in vacuum.[12] Natural titanomagnetite often intergrows with ilmenite

($FeTiO_3$). Atoms in the titanomagnetite lattice can also be replaced by other atoms Such as magnesium, aluminium, nickel, zinc, chromium, and vanadium.[1] Confocal Raman microscopy has the potential to provide compositional information with high spatial resolution and to reveal the key properties of titanomagnetite grains in rocks. To avoid laser induced alteration processes, we determined the laser power threshold for the oxidation of minerals from the titanomagnetite solid solution series. With the optimized settings, we have characterised single titanomagnetite grains in deep see basalts.

2. Materials and methods

To determine the laser power threshold we investigated synthetic magnetite (TM00) and titanomagnetites produced using the floating zone technique.[11] The titanium concentration of the synthetic samples ranged from 0 % to 86 %. Their titanium concentration was verified by thermomagnetic analysis.[2] To demonstrate compositional mapping we studied titanomagnetites in late Cretaceous sea floor basalt, drilled in the middle South Atlantic Ocean during the deep sea drilling project 73 (DSDP/ODP Leg 73, hole 524, core 33) in a water depth of 4796 m, 330 m below the sea floor.

Raman spectra were measured with a confocal Raman microscope (alpha300 R; WITec GmbH, Ulm, Germany). For excitation, we used the second-harmonic-generation (SHG, 532 nm, P_{max} = 22.5 mW) of a Nd:YAG laser, focused with a 100× objective (Nikon NA=0.90, 0.26 mm working distance). The objective was also applied for collection in backscattering geometry. The light was guided with a multimode optical fibre to the spectrometer, using the 50 µm diameter fibre core as a pinhole for the confocal setup. The diffraction-limited focus resulted in a lateral resolution of about 400 nm, and a focal depth of about 1 µm. The power distribution within the laser beam profile was Gaussian due to the TEM_{00} laser mode. Therefore, the energy input in the centre region was higher than at the rim of the laser spot.

A sharp edge filter rejected elastically scattered photons (Rayleigh scattering). Raman spectra were acquired with a lens-based spectrometer with the CCD-camera (1024 × 128 pixel, cooled to -65 °C) on each point with a spectral resolution of 3.51 cm^{-1} per CCD-pixel for the 600 mm^{-1} grating. The compositional maps were generated either by integrating the counts in the range of wavenumbers (sum filter) of characteristic Raman peaks or by using peak centre and peak width obtained by a Lorentzian fit.

The magnetic force microscope (MFM) was equipped with NanoScope IV controller (Veeco Metrology Inc., Santa Barbara, CA) using force sensors with a thin magnetic coating on the tip side of the cantilever (Multi75M-G, BudgetSensors, Sofia, Bulgaria). The sensors were magnetised shortly before use by placing them on one pole of a strong permanent magnet. For magnetic imaging, an atomic force microscope (AFM) was operated in tapping mode. The phase shift was recorded in lift mode after each topographic line scan (interleave mode).

An electron microprobe (cameca SX100) was used to measure the elemental composition of a similar titanomagnetite grain in the same thin section located a few millimetres away from the grain analysed by Raman spectroscopy. Operating parameters were 15 kV accelerating voltage, 20 nA beam current with a beam diameter of 1 µm. Counting rates of 30 s where used for Al, Ni, Ca, 100 s for Ti and 20 s for all other elements. Synthetic minerals and oxides were taken into account for calibration and a PAP (Pouchou and Pichoir) correction was applied to the raw data.[13]

3. Results

3.1 Laser induced oxidation

Reference spectra of the different titanomagnetites were obtained at a laser power of 1.5 mW. The Raman spectra were normalised to the A_{1g} peak and are shown in fig. 1. The Raman spectra did not depend on the orientation of the samples TM00 to TM86 (data not shown). Table 1 summarizes the Raman peak centres of the synthetic specimen as obtained by a Lorentzian fit. Confocal Raman microscopy on titanomagnetite is affected by laser absorption that leads to local heating. One can quantify the onset of oxidation with a series of Raman spectra measured with different laser intensities (fig. 2). The titanium content of the samples varied from $x = 0$ (TM00, magnetite, Fe_3O_4) to 86 % titanium (TM86, $Fe_{2.14}Ti_{0.86}O_4$). Each spectrum was integrated for 30 s. A progressive oxidation of titanomagnetite occurred as a function of the laser power. In an intermediate laser power range, the measured Raman spectra show features of both titanomagnetite and hematite. Both signatures were observed in the Raman spectra of the geologic titanomagnetite measured at a laser power of 14.9 mW (~1.1 MW/cm^2).

The threshold for initiating oxidation as a function of the titanium concentration was explored as illustrated in Fig. 3. A series of Raman spectra was taken by stepwise increasing laser power (see fig. 2). All spectra were recorded with a 30 s integration time. To detect spectral changes, the spectra were normalised to the A_{1g} peak, and then the spectrum taken at the previous power step was subtracted. Variations were considered significant if the differences exceeded 10% of the intensity in the range from 100 cm^{-1} to 3600 cm^{-1} (Fig. 3 inset).

Another important parameter is the irradiation time. To investigate the time dependence of the oxidation, a series of Raman spectra were recorded at the same position for the synthetic TM60 sample. The laser power was set at 2.2 mW (0.183 MW/cm^2), which is slightly below the threshold for oxidation at a 30 s irradiation time. Spectra were recorded continuously with an integration time of 30 s and normalised to the Rayleigh peak. After two minutes of irradiation, the spectrum began to show signs of alteration (fig. 4). Peaks at about 220 cm^{-1} and 290 cm^{-1} developed, which correspond to the spectral signature of hematite (A_{1g} and E_g respectively).

AFM and MFM measurements were taken after the laser irradiation to inspect whether the titanomagnetite surface was altered. Figure 5 shows the topography and the magnetic image of a locally oxidised surface of the natural titanomagnetite. The surface was irradiated for 5 s with laser powers of 5, 10, 15, and 20 mW respectively. Circular oxidation-craters with diameters of a few micrometers were generated (fig. 5(c)). The oxidation leads to a progressively deformed surface (fig. 5(a)). The same circular structure prevails in the MFM phase image (fig. 5(b)). Figures 5(d) and 5(e) show an oxidation crater directly after a laser irradiation with 20 mW. After 26 days, further changes of the deformed surface could be observed (fig. 5(f) and 5(g)). The small depression that was formed directly after laser irradiation in the centre (fig. 5(d)), likely underwent a process akin to an isostatic rebound 26 days later (fig. 5(f)).

3.2 Compositional mapping

Composition and structural characteristics of a titanomagnetite grain from an ocean floor basalt were determined using the Raman imaging procedure explained above. The optical micrograph revealed a striped pattern with varying reflectivity (Fig. 6(a)). The blue rectangle marks an area of 7×5 μm^2 (60 × 80 points) that was analysed with Raman spectroscopy using 10 s integration time at a laser power of 2 mW. The compositional map fig. 6(b) was calculated with a sum filter that integrates the counts from 210 cm^{-1} to 260 cm^{-1}. Sum filtering from 500 cm^{-1} to 580 cm^{-1} (fig. 6(d)) results in an inverted contrast. Two distinct materials were found as seen in the average spectra for the individual stripes (fig. 6(c)). The filter ranges for sum filtering are indicated by the light grey regions within the spectrum. The region used for the Lorentz fit of the A_{1g} peak is marked in dark grey. This peak was also taken into account for a more detailed analysis regarding peak position and peak width.

Lamellae with a width of a few micrometers in parallel two large cracks were observed in the optical micrograph (fig. 6(a)). In the following, we will refer to the material as "bright lamella" (red curve) and "dark lamella" (black curve) as they appearing in fig. 6(b). Average spectra of both regions were compared with reference spectra obtained on the synthetic samples (fig. 7). The inset outlines the A_{1g} Raman peak at about 660 cm^{-1}, which is characteristic for titanomagnetite. All spectra show the characteristic A_{1g} peak; other peaks change in position

(T_{2g}(1), from ~190 cm^{-1} to ~150 cm^{-1}), appear (T_{2g}(2), 480 cm^{-1}) or vanish (E_g, 310 cm^{-1}; T_{2g}(3), 540 cm^{-1} and the peak at 850 cm^{-1}) with increasing titanium content.

Figure 8 correlates the chemical information of the Raman spectra with information determined by AFM and MFM measurements on the geologic sample. The topography image in fig. 8(a) also reveals a striped pattern, which implies differences in material hardness. The magnetic image in fig. 8(b), show a clear stripe pattern alternating between magnetic and non-magnetic phases. The Raman map in fig. 8(c) was obtained from the peak centre position of a Lorentz fit on the A_{1g} peak and also shows these stripes. The Lorentz fit results in better contrast with decreased noise level as compared to the images calculated with the simple sum filter (fig. 6). All images clearly show the identical lamellar pattern. The structures in the MFM image correlate to the same structures as visible in the titanomagnetite distribution determined with Raman spectroscopy. An elemental microprobe analysis of a similar natural titanomagnetite revealed a slight exchange of magnesium against iron and a titanium and aluminium content that fluctuated (data not shown).

4 Discussion

4.1. Laser power

The alteration threshold decreases with increasing titanium concentration. Depending on the laser power, a partial oxidation could be observed within a certain range. Local temperatures towards the centre of the laser spot induced alteration whereas oxidation did not occur near the rim. In the spectrometer, however, scattered light from the centre and the surrounding region is averaged. When analysing titanomagnetites of unknown composition, the laser power should be limited to 1.5 mW (0.08 MW/cm^2) with integration times of several minutes. At laser powers <1.5 mW a safe mode of operation could be verified for the geologic and synthetic samples. Only spectra taken at laser powers below this threshold should be used for an uncompromised compositional analysis.

The results clearly show that laser-induced oxidation should be taken into account for a spectroscopic characterization of (titano-)magnetites. In addition, one also might adjust the oxygen fugacity during the measurement. Magnetite and titanomagnetite oxidise above ~240 °C in a normal atmosphere.[8, 14, 15] Raman measurements in an oxygen-free atmosphere might help reduce the tendency to oxidise. It should be mentioned, however, that noble gases can diffuse into synthetic magnetite above 500 °C during growth.[16] At about 600 °C, a phase transition from magnetite to other iron oxides can occur.[16-18] Such high temperatures may be achieved at high laser powers or strong focusing. For practical measurements under ambient conditions one has to balance measurement time and laser power to avoid alteration.

4.2. Deep sea basalt

With these optimized settings the composition of a natural titanomagnetite was mapped spectroscopically and two distinct phases was observed. A comparison with the reference spectra obtained on the synthetic specimen revealed subtle differences, which makes the identification of the mineral phases difficult. The peaks in the Raman spectra of the dark lamella (black curve) match best to the TM20 reference spectrum (fig. 7). The peak centre (665 cm^{-1}) of the A_{1g}, also better agrees with that of TM20 (664 cm^{-1}). The peak width (74 cm^{-1}), however, better matches to the width of the TM40 (76 cm^{-1}). The MFM measurements show that the dark lamella is a magnetic phase. Thus, it is reasonable to assume that the dark lamella correspond to a titanomagnetite with a titanium content around 20 % to 40 %.

The identification of the bright phase in fig. 6(b) is ambiguous. The relative intensities of the T_{2g}(2) and T_{2g}(3) peaks match to the TM40 spectrum. The peak width (66 cm^{-1}) is between the values for TM20 (54 cm^{-1}) and TM40 value. However, the peak at 235 cm^{-1} is missing in the reference spectra of the synthetic minerals. Also the A_{1g} peak position (673 cm^{-1}) does not fit well to the respective peak position of the synthetic titanomagnetites. These results indicate that the phase is most likely not a pure titanomagnetite phase but another iron oxide mineral. Various Raman spectra are available in the literature for chemically related iron oxides. For example, Faria et al. reported similar Raman spectra for a laser heated wüstite, where decomposition leads to α-Fe and Fe_3O_4, and assumed that Fe_3O_4 has been transformed into hematite (α-Fe_2O_3).[14] Raman spectra of maghemite (γ-Fe_2O_3), which can be seen as iron-deficient form of magnetite, also coincide with that of the bright lamella.[19-21] Ilmenite is another candidate because the peaks at 235 cm^{-1} bright lamella and 229 cm^{-1} ilmenite match. However, there are discrepancies for the A_{1g} peak position (680 cm^{-1} for ilmenite) and in particular the peak at ~480 cm^{-1}, which is not present in ilmenite. The bright lamella is not ferromagnetic because in MFM the bright lamella does not show a signal (fig. 8(b)). Titanomagnetites with x = ~0.75 are paramagnetic at room temperature, therefore such a titanomagnetite phase cannot be excluded with the MFM measurements. However maghemite and ilmenite were disproved with MFM, given that both phases are magnetic.

Microprobe data provides additional insights into the composition of the mineral phases. Also the elemental composition showed an alternating exchange of ions. The variations may be caused by growth-zoning in natural magnetite. It is reasonable to assume that the variation observed by the microprobe measurement also affects the Raman spectra. The exchange of iron and magnesium ions could explain the deviation of the spectrum of the bright lamella from a pure mineral phase. The composition of that phase, however, remains an open question to further study.

5. Conclusion

Titanomagnetite Raman spectra vary in peak position, intensity and width as a function of titanium concentration and degree of oxidation. Hence, Raman spectroscopy can identify the composition and oxidation state of minerals of the magnetite-ülvospinel solid solution series. Depending on the titanium content, the mineral starts to oxidise above a certain laser power threshold. Adjusting the laser power just below the threshold for immediate oxidation is insufficient because time-dependent oxidation processes can still occur. The laser heating causes an altered spot on the surface whose diameter and height depends on the laser power and irradiation time. Oxidation continues weeks after laser application. Laser power should be maintained well under the threshold to avoid oxidation.

Based on these findings, we investigated a natural basalt collected from the deep ocean floor. Compositional Raman mapping revealed a lamellar structure. One mineral phase can be attributed to a titanomagnetite with a titanium content of about 20 % - 40 %. The other phase could not be unambiguously identified.

Acknowledgements

We thank Gloria Almonacid Caballer for fruitful discussions and the Deutsche Forschungsgemeinschaft (DFG) for funding this project under grant STA 1026/2-1

References

[1] A. Wang, K. E. Kuebler, B. L. Jolliff, L. A. Haskin, *American Mineralogist* **2004**, *89*, 665.
[2] S. A. Gilder, M. Le Goff, *Geophys. Res. Lett.* **2008**, *35*.
[3] M. Marshall, A. Cox, *Nature* **1971**, *230*, 28.
[4] J. Verhoogen, *The Journal of Geology* **1962**, *70*, 168.
[5] L. Tatsumi-Petrochilos, S. A. Gilder, P. Zinin, J. E. Hammer, M. D. Fuller, *AGU Fall Meeting Abstracts* **2008**, A791+.
[6] A. Ellery, D. Wynn-Williams, *Astrobiology* **2003**, *3*, 565.
[7] J. Popp, M. Schmitt, *Journal of Raman Spectroscopy* **2004**, *35*, 429.
[8] O. N. Shebanova, P. Lazor, *Journal of Raman Spectroscopy* **2003**, *34*, 845.
[9] H. C. Soffel, E. Appel, *Physics of The Earth and Planetary Interiors* **1982**, *30*, 348.
[10] L. V. Gasparov, D. B. Tanner, D. B. Romero, H. Berger, G. Margaritondo, L. Forro, *Physical Review B* **2000**, *62*, 7939.
[11] B. J. Wanamaker, B. M. Moskowitz, *Geophys. Res. Lett.* **1994**, *21*.
[12] M. Lewis, *Geophysical Journal of the Royal Astronomical Society* **1968**, *16*, 295.
[13] J. L. Pouchou, F. Pichoir, *La Recherche Aérospatiale* **1984**, *3*, 13.
[14] D. L. A. d. Faria, S. V. Silva, M. T. d. Oliveira, *Journal of Raman Spectroscopy* **1997**, *28*, 873.
[15] B.-K. Kim, J. A. Szpunar, *Scripta Materialia* **2001**, *44*, 2605.
[16] T. Matsumoto, K. Maruo, A. Tsuchiyama, J.-I. Matsuda, *Earth and Planetary Science Letters* **1996**, *141*, 315.
[17] D. P. Burke, R. L. Higginson, *Scripta Materialia* **2000**, *42*, 277.
[18] J. Tominaga, K. Wakimoto, T. Mori, M. Murakami, T. Yoshimura, *Transactions of the Iron and Steel Institute of Japan* **1982**, *22*, 646.
[19] I. Chamritski, G. Burns, *The Journal of Physical Chemistry B* **2005**, *109*, 4965.
[20] J. C. Rubim, M. H. Sousa, J. C. O. Silva, F. A. Tourinho, *Brazilian Journal of Physics* **2001**, *31*, 402.
[21] M. H. Sousa, F. A. Tourinho, J. C. Rubim, *Journal of Raman Spectroscopy* **2000**, *31*, 185.

Tables

Table 1: Raman peak centres of the synthetic specimen as obtained by a Lorentzian fit together with literature values for magnetite. [a] A. Wang et al., *Am. Miner.* **2004**, *89*, 665.

Ti/(Fe+Ti) Mode	Raman peak centre (cm^{-1})				
	$T_{2g}(1)$	E_g	$T_{2g}(2)$	$T_{2g}(3)$	A_{1g}
TM00[a]	192	306	-	538	668
TM00	197.8	304.4	-	536.8	664.9
TM20	193.2	309.0	-	540.7	664.1
TM40	155.8	309.2	455.1	549.7	647.5
TM60	155.7	318.2	464.2	-	647.5
TM86	154.4	-	477.7	-	654.3

Figures

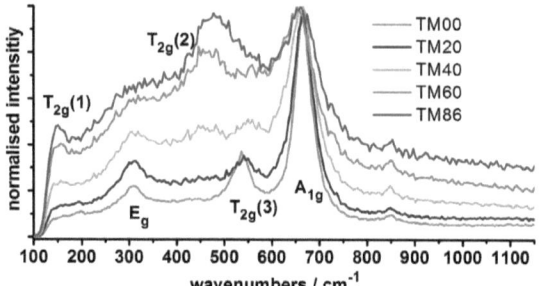

Fig. 1. Raman spectra of synthetic titanomagnetite normalized to the A_{1g} peak.

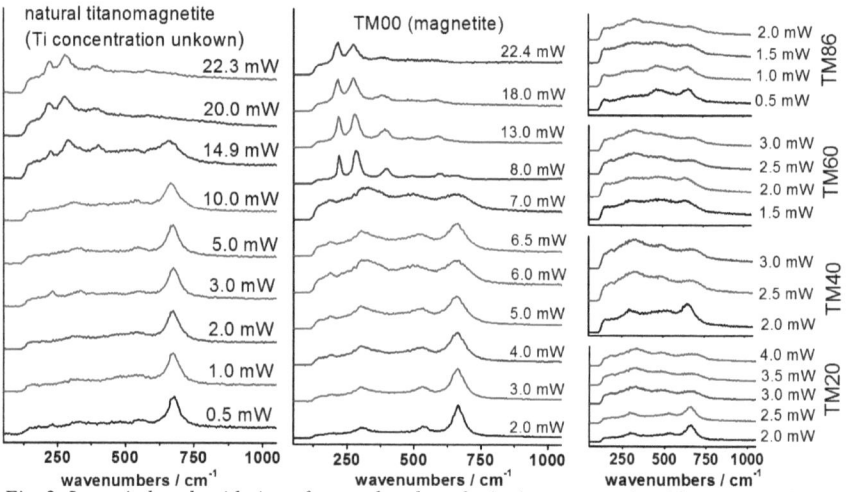

Fig. 2. Laser induced oxidation of natural and synthetic titanomagnetite. Above a certain laser power threshold the Raman spectra a hematite signature occurred.

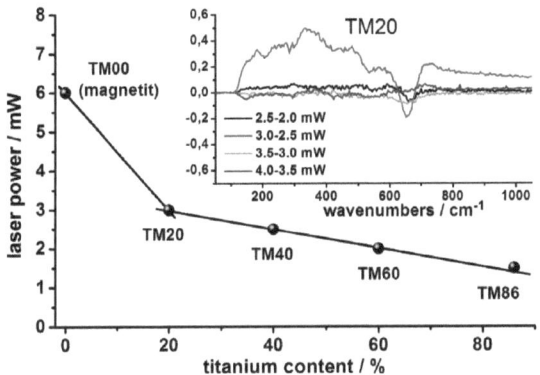

Fig. 3. Laser power threshold for the oxidation of titanomagnetite. The inset shows spectra differences.

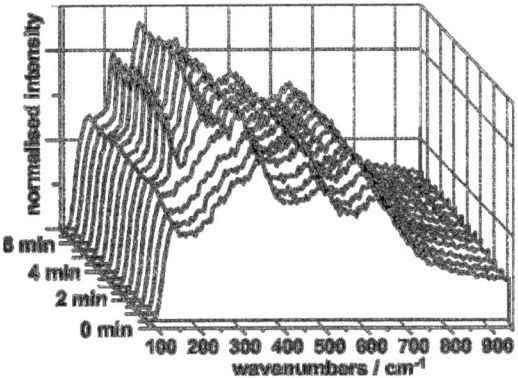

Fig. 4. Time dependence of the laser induced oxidation of TM60. The spectra were taken with a laser power of 2.2 mW. After two minutes of irradiation hematite peaks occurred at 220 cm^{-1} (A_{1g}) and 290 cm^{-1} (E_g).

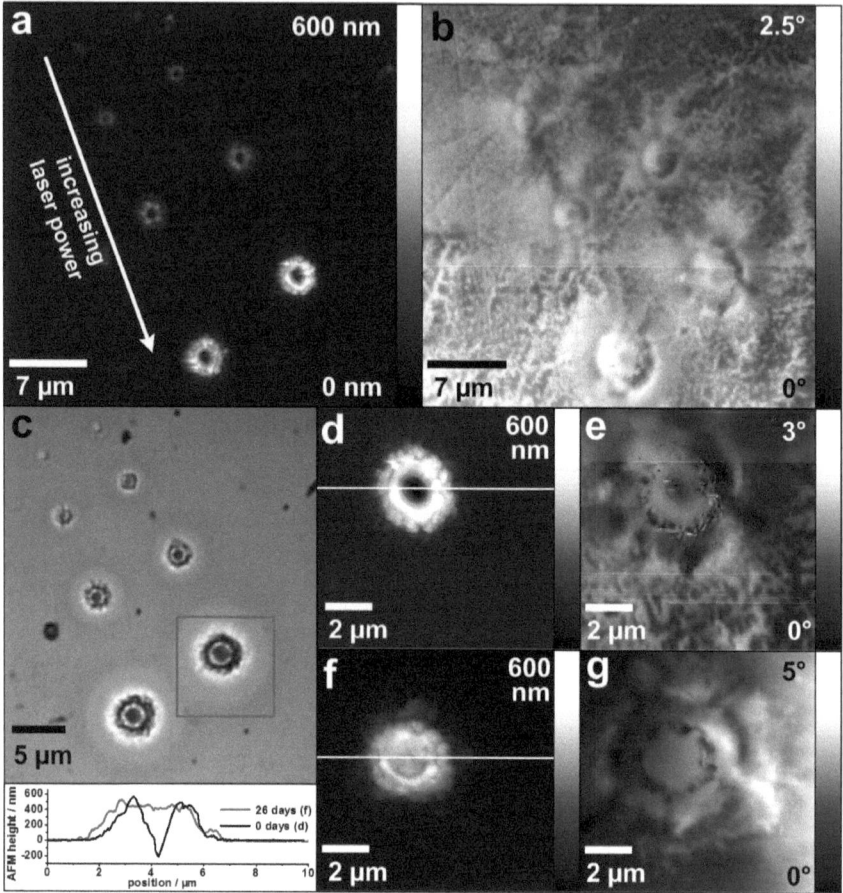

Fig. 5. MFM and AFM images of a laser irradiated natural titanomagnetite surface. Oxidation was achieved by means of 5 s laser irradiation with 5, 10, 15 and 20 mW (top to bottom). (a) topography, (b) corresponding phase image (lift height 100 nm) and (c) optical micrograph. (d,e) Topography and MFM images of the oxidation spot generated with 20 mW were measured directly after the laser irradiation (lift height 40 nm). (f,g) The same spot after 26 days storage at ambient lab conditions. The inset at the lower left shows the topographic cross section of (d) and (f). The central depression disappeared after 26 days.

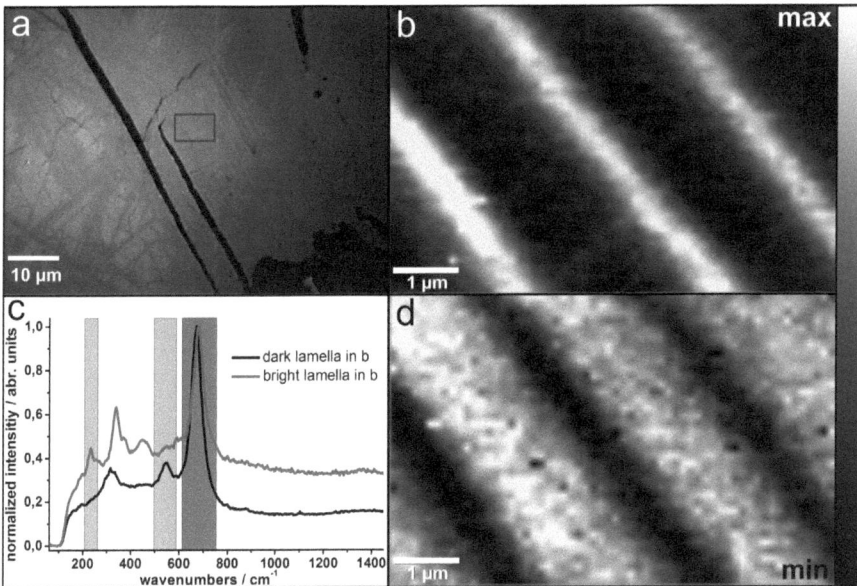

Fig. 6. Compositional mapping of a natural titanomagnetite grain from a mid-oceanic ridge basalt. (a) Optical micrograph. The region of the Raman maps was located close to a shrinkage crack and is marked with a blue rectangle. (b) Compositional map calculated with a sum filter from 210 cm^{-1} - 260 cm^{-1}. (c) Average spectra obtained from both types of stripes. The red curve is an average over the bright lamella in (b), the black curve over the dark ones. (d) Compositional map calculated with sum filter from 500 cm^{-1} - 580 cm^{-1}. Both sum filter regions were marked in light grey, the region used for the Lorentz fit of the A_{1g} peak is marked in dark grey.

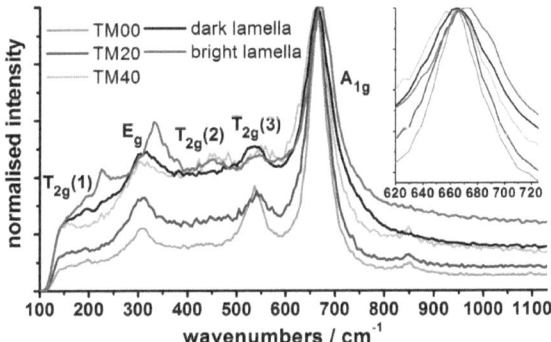

Fig 7: Reference spectra of synthetic titanomagnetite and average spectra taken on the geologic sample. All spectra were normalized to the A_{1g}. The inset shows the A_{1g} peak in detail.

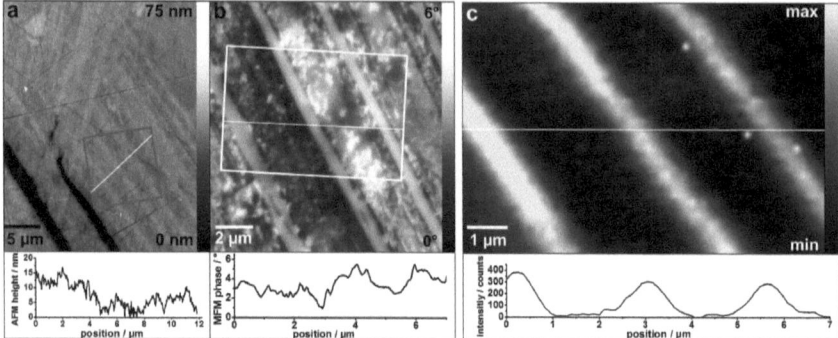

Fig. 8: AFM, MFM and Raman maps of a titanomagnetite from a sea floor basalt. (a) topographic overview. (b) The MFM phase shift with 50 nm lift height was taken in the region of interest indicated in (a). (c) Raman map acquired in the region indicated in (b). The Raman image was derived from the peak centre position of the A_{1g} peak determined with a Lorentz fit.

Acknowledgements

There are many people I would like to thank for their support that made this dissertation possible. First, I thank PD Dr. Robert Stark, who gave me the opportunity to do research in his group and laboratories. With his guidance and encouragement on individual topics and allowing me the freedom to work on my own subjects of interest, I was able to create this thesis.

I am also grateful to Prof. Wolfgang Heckl, who let me start in his group, and to Prof. Stefan Sotier, who made the contact.

Many thanks also go to Dr. Alexander Gigler for his assistance during the measurements and the interpretation of the results. I am also grateful for the support of the other members in the group (Marek Janko, Dr. Marc Hennemeyer, and Polina Davydovskaya) and the entire parallel STM group of Prof. Markus Lackinger. I enjoyed working together and discussing everything and, of course, having activities outside of my scientific work.

In addition, I thank my colleagues in the Laser Discharge Unit of the former Tui-Laser AG, in particular, Dr. Sebastian Spörlein, Dr. Claus Strowitzki, Alexandar Djordevic, Dr. Andreas Görtler and Ansgar Mattern for giving me the chance to work on the problems of deteriorating excimer optics. I thank them not only for the experience with excimer laser systems and related technologies but also for their excellent teamwork. It has been a great pleasure working with them for all these years.

Last but not least, I thank my parents and my brother, who have supported me throughout the preparation of this dissertation.

I want morebooks!

Buy your books fast and straightforward online - at one of world's fastest growing online book stores! Environmentally sound due to Print-on-Demand technologies.

Buy your books online at
www.morebooks.shop

Kaufen Sie Ihre Bücher schnell und unkompliziert online – auf einer der am schnellsten wachsenden Buchhandelsplattformen weltweit! Dank Print-On-Demand umwelt- und ressourcenschonend produziert.

Bücher schneller online kaufen
www.morebooks.shop

KS OmniScriptum Publishing
Brivibas gatve 197
LV-1039 Riga, Latvia
Telefax: +371 686 204 55

info@omniscriptum.com
www.omniscriptum.com

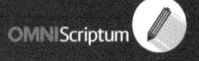

Printed by Books on Demand GmbH, Norderstedt / Germany